HISTORIÆ NATURALIS CLASSICA, XXVIII

HISTORIÆ NATURALIS CLASSICA

EDIDERUNT

J. CRAMER ET H. K. SWANN

TOMUS XXVIII

THE
BRITISH DESMIDIEAE

BY

JOHN RALFS

REPRINTS OF LATER STARTING POINT BOOKS
FOR BOTANICAL NOMENCLATURE, VOLUME 4

REPRINT 1962

BY J. CRAMER · WEINHEIM

WHELDON & WESLEY, LTD AND HAFNER PUBLISHING CO.
CODICOTE, HERTS.   NEW YORK, N.Y.

THE
BRITISH DESMIDIEAE

BY

JOHN RALFS

REPRINT 1962

BY J. CRAMER · WEINHEIM

WHELDON & WESLEY, LTD AND HAFNER PUBLISHING CO.
CODICOTE, HERTS.   NEW YORK, N.Y.

COPYRIGHT OF THE SERIES
BY J. CRAMER, WEINHEIM

PRINTED IN GERMANY

THE

BRITISH DESMIDIEÆ.

BY

JOHN RALFS, M.R.C.S.,
HONORARY MEMBER OF THE PENZANCE NATURAL HISTORY SOCIETY, ETC.

THE

DRAWINGS

BY

EDWARD JENNER, A.L.S.

LONDON:
REEVE, BENHAM, AND REEVE,
KING WILLIAM STREET, STRAND.
1848.

PRINTED BY RICHARD AND JOHN E. TAYLOR,
RED LION COURT, FLEET STREET.

TO

WILLIAM BORRER, ESQ.,

F.R.S., L.S. &c.

DEAR SIR,

As your suggestion induced me to study the Microscopic Algæ with greater diligence, and as I have been mainly indebted to your assistance and encouragement for the successful completion of the present Monograph, I feel it a grateful duty to dedicate to you the collected fruits of my investigations, and, sincerely thanking you for the kind permission to associate them with a name so honourably distinguished in Botanical Science,

I am, dear Sir,

Your truly obliged,

JOHN RALFS.

PREFACE.

Although this work has greatly exceeded the limits which I originally contemplated, I am very far from thinking that it has exhausted the subject. About one-third of the Desmidieæ which it describes have been discovered in this country since the prospectus was issued; a fact which may well warrant the expectation that a rich harvest yet remains to reward the diligence of future labourers. In regard also to the reproductive organs; whilst I have the pleasure of noticing that the sporangia of fifty species are here figured, most of them for the first time, I would point out that those of a much larger number are still unknown, and that every spring and summer many are brought to light.

In compliance with the request of some of my Subscribers, I have added a few foreign habitats, which will give some idea of the geographical distribution of the species. This I am enabled to do through the kindness of M. de Brébisson and Professor Bailey, who have furnished me with complete lists of the species found by them. It must of course be considered merely the first rude chart of their range, and I earnestly deprecate any comparison of it with what Professor Harvey has accomplished for the Marine Algæ in his splendid 'Phycologia Britannica': nevertheless enough is here exhibited to show the cosmopolite character of these plants, since not only has almost every British species been found at

Falaise by M. de Brébisson, but most of them have also American habitats.

It has been my desire throughout to write impartially, and I am not aware that I have, in a single instance, neglected the claim of priority, or appropriated to myself the discoveries of others. Should any blemishes of this kind appear, they will, I trust, be attributed to the circumstance that this work has been composed at a distance from the metropolis, and without access to many works which it would have been desirable to consult. Amongst these I have more especially to regret that I have not seen some valuable papers on the Desmidieæ by Ehrenberg, Corda, and Morren, which have been published in various foreign periodicals.

In these pages the name appended to a species merely indicates the author of the specific name, and has no reference to its genus. This departure from the usual custom seems to me to require no vindication; for it is surely unjust that the credit due to the discoverer or first describer of a plant should be ascribed to one whose sole merit in regard to it has been to transfer it from one genus to another. To those who prefer the common method, the synonyms will afford the requisite information.

I consider *Merismopedia* and *Trochiscia* to belong to the Palmelleæ, and have therefore omitted them, although Kützing and Meneghini include them in this family. I am also unacquainted with any species of *Sphærastrum* which can rank with the Desmidieæ.

Penium Digitus, and, except in a few cases which are pointed out, the species of *Closterium*, *Docidium* and *Micrasterias*, are, on account of their comparatively large size, magnified only 200 diameters, but all the others are magnified 400 diameters.

Mr. Jenner and myself have used Mr. Ross's microscopes, and I think it due to that skilful optician to state, that for clearness of vision we have never seen them surpassed.

I have added, as an Appendix, a list of as many foreign species as have come to my knowledge; and, although I fear that a few of Ehrenberg's may have been omitted, I have some confidence that this work describes and figures a greater number of species than have before been systematically arranged. M. de Brébisson and Professor Bailey have kindly permitted me to insert species which they have recently detected, and which are now first published, and also to make use of their drawings. The last plate has been reserved for these drawings and for others which Professor Kützing has presented me of species which he has described but not figured. I have to regret that the large size of the *Closteria* did not allow the introduction of figures of several interesting foreign species.

The drawings of the British species, with very few exceptions, were made by Mr. Jenner; but they have, in almost every instance, been carefully re-examined by myself and compared with the plants, and the engravings have been executed under my own superintendence. From the commencement of this undertaking Mr. Jenner's help has been invaluable. Not only has he taken the greatest pains to ensure the accuracy of the drawings and furnished me with copious notes, but to his exertions I owe a very large proportion of my Subscribers. Indeed I can truly say that he has shown as great an interest in the work as if it were his own, and exerted himself as much for me as he would have done for himself.

To the kind recommendations of Mr. Bowerbank, who is

as distinguished for his indefatigable and successful microscopic observations as for his constant readiness to encourage his fellow-labourers in Natural History, I owe the patronage of many of the eminent naturalists whose names adorn the list of my Subscribers; and, to render this Monograph more complete and deserving of such support, he has taken the trouble of measuring all the species. Tedious and uninviting as the task must have been, these accurate measurements will show how much his love of science has contributed to enrich a work of which I regard them as one of the most important features.

It gives me sincere pleasure to acknowledge how much I am indebted to the kindness of M. de Brébisson, who has so long studied these Algæ, and who is deservedly esteemed the highest authority on the subject. By his valuable aid my nomenclature has been rendered far more perfect than it could otherwise have been. He has most liberally supplied me with specimens and drawings of almost every species; has copied, for my use, figures contained in works which I had no opportunity of consulting, and has at all times readily furnished me with such information as I required. To the extent indeed of my obligations in these respects every page of this work bears frequent but inadequate testimony.

Professors Kützing and Bailey have also afforded me important assistance, by supplying me with drawings of many species and by the communication of interesting facts.

My friends, the Rev. M. J. Berkeley, Dr. Dickie, Mr. Sidebotham, and Mr. Thwaites, have materially aided me in the execution of this Monograph by sending me their observations and specimens. Dr. Williams of Swansea has obligingly furnished me with copious translations from German works.

To no one are my obligations greater than to Mr. Borrer: by his advice and assistance the work has been freed from many defects that would have impaired the utility which I trust it may now possess. I gladly return my thanks also to my valued friend the Rev. H. Penneck, who has afforded me his assistance throughout its progress and in its careful preparation for the press.

My thanks are due likewise to many other friends, for the interest which they have taken in this publication, and for their exertions to secure me from pecuniary risk; especially to Mrs. Griffiths, Miss Warren, Professor Allman, Mr. Andrews, Professor Balfour, Mr. Coates, Mr. R. Q. Couch, Mr. Dillwyn, Mr. P. Grant, Dr. Greville, Mr. Gutch, Mr. W. Hanson, Professor Harvey, Rev. D. Landsborough, Mr. Moggridge, Mr. Moore, Dr. Shapter, Mr. W. Thompson, Mr. Topping, Mr. Tudor, and Mr. Williamson.

I have to apologize for the long delay in publishing. It has been caused partly by ill health, but chiefly by the continual discovery of additional matter and my wish that no labours on my part should be wanting. When however I state that I have thus been enabled to add the descriptions of a large proportion of the species and of one-half of the sporangia, besides the Appendix of foreign species, and at least ten plates, I may be permitted to hope that the Subscribers will readily excuse a delay which has so materially increased the extent of the work without increasing its price.

SUBSCRIBERS.

Royal College of Surgeons of England.
Royal Dublin Society.
Royal Society of Edinburgh.
University and King's College, Aberdeen.
Royal Geological Society of Cornwall.
Botanical Society of Edinburgh.
Microscopical Society of London.
Microscopical Society of Dublin.
Microscopical Society of Bristol.
Belfast Library.
Trinity College Library, Dublin.
Devon and Exeter Botanical and Horticultural Society.
Devon and Exeter Institution.
Royal Horticultural Society, Cornwall.
University Library, Edinburgh.
Wernerian Natural History Society of Edinburgh.
Adam, Thomas, Esq., Halifax.
Allcard, John, Esq., F.L.S., &c., 65, Lombard Street, London.
Allingham, Mr. William, Reigate.
Allman, G. J., M.B., M.R.I.A., F.R.C.S.I., M.B.S.E., &c., Professor of Botany in Trinity College, Dublin.
Anderson, T. P., Esq., M.R.I., 45, Harewood Square, London.
Andrews, W., Esq., M.R.I.A., B.S.L., Secretary to the Natural History Society of Dublin. 2 *copies*.
Ansted, D. T., Esq., M.A., F.R.S., Professor of Geology, King's College, London.
Arnott, George A. Walker, Esq., LL.D., F.R.S.E., F.L.S., Regius Professor of Botany, Glasgow.
Ayerst, Mr. Thomas, jun., Hill Cottage, Newenden, Kent.
Babington, C. C., Esq., M.A., F.L.S., G.S., &c., St. John's College, Cambridge.
Babington, Churchill, Esq., B.A., Fellow of St. John's College, Cambridge.
Bagster, George, Esq., Paternoster Row.
Baillie, Mrs., Hill Park, Westerham, Kent.
Balfour, J. Hutton, M.D., F.R.S.E., L.S., &c., Regius Professor of Botany, Edinburgh.
Ball, Robert, Esq., M.R.I.A., B.S.E., Vice-President of the Geological Society of Dublin.

Bardsley, James L., M.D., F.L.S., Manchester.
Barrow, R., Esq., M.R.C.S., Manchester.
Bass, Mrs., Brighton.
Bayne, A., Esq., Tunbridge Wells.
Beaufort, Rev. D. A., 11, Gloucester Place, Portman Square.
Beck, Mr., Coleman Street, City.
Bell, Frederick J., Esq., Woodlands, near Maldon, Essex.
Bell, Robert J., Esq., M.R.C.S., Mickleover House, near Derby.
Bell, Thomas, Esq., F.R.S., L.S., G.S., &c., Professor of Zoology, King's College, London.
Bennett, W., Esq., 7, Park Village West, Regent's Park.
Bentley, Robert, Esq., King's College, London.
Bergin, Thomas F., Esq., M.R.I.A., President of the Microscopical Society of Dublin.
Berkeley, Rev. M. J., M.A., F.L.S., King's Cliffe.   3 *copies*.
Bird, Golding, M.D., M.A., F.R.S., F.L.S., Lecturer on Botany, Guy's Hospital, &c., Myddelton Square, London.
Birkby, Henry, Esq., Rochdale.
Birkett, John, Esq., F.L.S., 2, Broad Street Buildings.
Bladon, James, Esq., Pont-y-Pool, Monmouthshire.
Bond, Mr., Richards Buildings, Hackney Road.
Borrer, W., Esq., F.R.S., L.S., M.B.S.E., &c., Henfield.   2 *copies*.
Bowerbank, J. S., Esq., F.R.S., L.S., G.S., &c., Highbury Grove, London.
Boyle, W. A., Esq., F.L.S., 37, Granville Square, Pentonville.
Braikenridge, G. W., Esq., F.S.A. & G.S., Broomwell House, Brislington, near Bristol.
Briggs, Mr. Charles, Brighton.
Bright, Richard, M.D., F.R.S., Saville Row, London.
Brightwell, Thomas, Esq., F.L.S., Norwich.
Brittain, Thomas, Esq., Manchester.
Broome, C. E., Esq., M.A., M.B.S.L., B.S.E., &c., Clifton.
Broomfield, William Arnold, M.D., F.L.S., Ryde, Isle of Wight.
Brown, Edwin, Esq., Burton-on-Trent.
Brown, H., Esq., Lewes.
Brown, Isaac, Esq., M.B.S.E., Dorking.
Brown, Robert, Esq., D.C.L., F.R.S., V.P.L.S., &c., British Museum.
Browne, Henry, M.D., M.R.C.S., Manchester.
Buckland, The Very Rev. W., D.D., Dean of Westminster, F.R.S., G.S., F.L.S., &c., Professor of Geology and Mineralogy, Oxford.
Buffham, Mr. William, Rochdale.
Bunting, Frederick, Esq., Westhall Place, Cheltenham.
Busk, George, Esq., F.L.S., F.R.C.S., Croom's Hill, Greenwich.
Button, Charles, Esq., M.M.S.L., Holborn.
Callwell, Robert, Esq., M.R.I.A., Dublin.
Capron, J. R., Esq., Guildford.
Carey, J., Esq., Strand, London.

SUBSCRIBERS.

Carpenter, W. B., M.D., F.R.S., 6, Regent's Park Terrace, Gloucester Gate.
Carpenter and Westley, 24, Regent Street, London.
Casey, J., Esq., Lea Bridge Road, London.
Cattell, Mr. J., Westerham, Kent.
Clark, Rev. W., M.D., F.R.S., Professor of Anatomy, Cambridge.
Clark, Edwin, Esq., George Street, Westminster.
Coates, John, Esq., M.R.C.S., &c., Rochdale.
Coathupe, Ch. Thornton, Esq., Wraxall, Bristol.
Cobb, Rev. S. W., Ightham, Kent.
Colbran, Mr. J., Tunbridge Wells.
Collins, Henry, Esq., Berkeley Lodge, Chichester.
Combe, Miss M. A., Cobham Park, Surrey.
Cook, George, Esq., Manchester.
Couch, R. Q., Esq., M.R.C.S., M.R.C.G.S., &c., Penzance.
Cullen, W. H., M.D., Sidmouth.
Curme, George, Esq., M.R.C.S., Dorchester.
Currer, Miss Richardson, Eshton Hall, Gargrave, Yorkshire. 2 *copies*.
Cutler, Miss, Budleigh Salterton, Devonshire.
Dalrymple, J., Esq., F.R.C.S., Grosvenor Street, London.
Daly, Rev. W. J., M.A., O.M.I., Penzance.
Dancer, J. B., Esq., Manchester.
Darby, Alfred, Esq., Coalbrook-dale.
Darker, W. Hill, Esq., Paradise Street, Lambeth.
Daubeny, Charles G. B., M.D., F.R.S., L.S. and G.S., Hon. M.R.I.A., Regius Professor of Botany, Chemistry, and Rural Economy, Oxford.
Davis, Mr. John, Mile-end Road, London.
Davis, Richard, Esq., F.L.S., H.S., and R.B.S., 9, St. Helen's Place.
Deane, H., Esq., Clapham Common.
Death, Mr. W., Petworth.
De Boos, E., Esq., Elder Street, London.
De la Rue, Warren, Esq., Bunhill Row, London.
De Ruvignes, C. H., Esq., Cape of Good Hope.
Diamond, Mr. A. G., Brenchley, Kent.
Diamond, Hugh W., Esq., Frith Street, Soho.
Dickenson, Henry, Esq., Coalbrook-dale.
Dickie, G., M.D., M.B.S.E., Lecturer on Botany, Aberdeen.
Dickinson, J., Esq., M.A., M.D., F.L.S., &c., Lecturer on Botany, Liverpool.
Dillwyn, L. W., Esq., F.R.S., L.S., &c., Sketty Hall, Swansea. 2 *copies*.
Donald, Ernest, Esq., Aberdeen.
Doxat, Alexis James, Esq., Putney Heath.
Drummond, J. L., M.D., Professor of Anatomy and Physiology, Royal Belfast Institution.
Durrant, Richard, Esq., London.
Elliott, W. H., M.D., F.L.S., M.R.C.S., &c., Boverie House, Mount Radford, Exeter.
Elsey, J. R., Esq., Lonsdale Square, Islington.

Estlin, J. B., Esq., F.L.S., F.R.C.S., Bristol.
Etheridge, Mr., Cheltenham.
Ferris, Octavius Allen, Esq., Victoria Park, Manchester.
Fielding, H. B., Esq., F.L.S., Bolton Lodge, Lancaster.
Foord, George, Esq., Walter Street, Soho. 2 *copies*.
Foot, Simon, Esq., M.R.I.A., B.S.E., &c., Holly Park, Dublin.
Forbes, Edward, Esq., F.R.S., L.S., G.S., &c., Professor of Botany, King's College, London.
Ford, John William, Esq., 9, Duke Street, London Bridge.
Forster, Edward, Esq., F.R.S., Treas. & V.P.L.S., &c., Woodford.
Freeman, J. D., Esq., Plymouth.
Fry, P. W., Esq., Montague Street, Russell Square.
Gamwell, F., Esq., Myddelton Terrace, Islington.
Gillett, W. S., Esq., Wilton Street, Grosvenor Place.
Godley, Mr. William, Wallingford, Berks.
Goodsir, J., Esq., F.R.S., Professor of Anatomy, University, Edinburgh.
Gordon, P. S., Esq., Craigmyle.
Gorham, J., Esq., M.R.C.S., Tunbridge.
Graham, R., Esq., Buxton House, Walthamstow.
Grant, J. W., Esq., Calcutta.
Gratton, Joseph, Esq., 94, Shoreditch, London.
Graves, Mr. George, Hackney.
Gray, T., Esq., Manchester.
Gray, John, Esq., Greenock.
Greenwood, Alfred, Esq., M.B.S.E., Chelmsford.
Greville, R. K., LL.D., F.R.S.E., &c., Edinburgh.
Griffith, Mrs., Torquay.
Grindon, L. H., Esq., Manchester.
Grosvenor, Rt. Hon. Lord Robert, M.P., 107, Park Street, Grosvenor Square.
Gutch, J. W. G., Esq., M.R.C.S., B.S.L., &c., Great Portland Street.
Gwilt, George, Esq., Union Street, Borough.
Hair, Quintin, B., Esq., 288, Regent Street, London.
Hamilton, Edward, M.D., F.L.S., &c., 22, Grafton Street, Bond Street.
Hankey, J. A., Esq., F.L.S., 8, Grosvenor Square, London.
Hanson, Mr. William, A.L.S., Reigate.
Harvey, W. H., M.D., M.R.I.A., Professor of Botany to the Royal Dublin Society, Trinity College, Dublin. 2 *copies*.
Hatch, Mr. Benjamin, Tenterden.
Heddle, Robert, Esq., Orkney.
Henfrey, A., Esq., F.L.S., Lecturer on Botany, Middlesex Hospital, 6, Shrewsbury Cottages, Holland Road, North Brixton.
Henslow, Rev. J. S., M.A., F.L.S., G.S., C.P.S., Professor of Botany, Cambridge, and Examiner in Botany in the University of London.
Heryer, Jacob, Esq., Great St. Helens, Bishopgate Street.
Hetling, George, Esq., Bristol.
Heyworth, Rev. J., Henbury Hill, near Bristol.
Holmes, George B., Esq., Horsham, Sussex.

SUBSCRIBERS.

Hooker, Joseph Dalton, M.D., F.L.S., Mem. Imp. Acad. Cæs. Leop. Nat. Curiosorum, &c.
Hooker, Sir W. J., K.H., LL.D., F.R.S., V.P.L.S., &c., Kew.
Hore, Rev. W. S., M.A., F.L.S. and G.S., Stoke, Devonport.
Howard, D., Esq., 22, Berners Street, Oxford Street.
Howman, J. A. Knightly, Esq., Brompton Square.
Hurst, Z. D., Esq., F.G.S., Aylesbury.
Ingpen, Abel, Esq., A.L.S., King's Road, Chelsea.
Inkson, Mr. William, Chichester.
Irving, General, Solway, Kirkcudbright.
Jackson, John, Esq., Hackney.
Jackson, Rev. W., Fellow of Worcester College, Oxford, and Incumbent of All-Saints, Sidmouth.
Janson, Henry U., Esq., Pennsylvania, Exeter.
Janson, J., Esq., F.L.S., Stoke Newington.
Jenyns, Rev. Leonard, M.A., F.L.S., G.S., and C.P.S., Swaffham Bulbeck, Newmarket.
Johnson, Christopher, jun., Esq., M.R.C.S., Lancaster.
Johnston, G., M.D., LL.D., F.R.C.S.E., Berwick-upon-Tweed.
Jones, S. M., Esq., Sloane Street, Chelsea.
Jones, Capt. Theobald, Royal Navy, M.P., F.L.S., G.S., &c., 30, Charles Street, St. James's. *2 copies.*
Josephs, Walter, Esq., 15, Crown Street, Finsbury, London.
Just, J., Esq., M.B.S.E., Grammar School, Bury, Lancashire.
Keddie, W., Esq., Glasgow.
Kennedy, Benjamin, Esq., F.L.S., Clapton, Middlesex.
King, John, Esq., Ipswich.
Kirkpatrick, George, Esq., Isle of Wight.
Langford, John, Esq., Lewes.
Leach, G., Esq., F.Z.S., Melina Place, St. John's Wood.
Lealand, Mr., Clarendon Street, Somers' Town.
Lear, Henry, Esq., Arundel.
Leeson, H. B., M.D., Greenwich.
Leighton, Rev. W. A., B.A., M.B.S.E., Luciefelde, Shrewsbury.
Lemon, Sir Charles, Bart., M.P., F.R.S., G.S., H.S., President of R.G.S. of Cornwall, &c.
Leonard, S. W., Esq., M.M.S.L., 83, Upper Stamford Street.
Leyland, R., Esq., Halifax, Yorkshire.
Lighton, Rev. Sir Christopher R., Bart., Epsom.
Lighton, Thomas, Esq., Clifton.
Lindley, John, Ph.D., F.R.S., L.S., H.S., Professor of Botany, University College, London, &c.
Lister, J. J., Esq., F.R.S., &c., Upton, Essex.
Lower, M. A., Esq., F.A.S., Lewes.
Marshall, Matthew, Esq., Bank of England.
Martin, Major, Ardrossan, Ayrshire.
Martin, Thomas, Esq. M.R.C.S., Reigate.

Massey, Mrs., 22, Queen Street, May Fair, London.
Maughan, James, Esq., Liverpool.
May, Charles, Esq., Ipswich.
Melville, A. Gordon, M.D., M.R.C.S., British Museum.
Milner, W. R., Esq., M.R.C.S., Wakefield.
Mitten, Mr. William, A.L.S., Hurstpierpoint.
Moggridge, M., Esq., M.B.S.L., Swansea.
Moore, D., Esq., M.R.I.A., B.S.L., &c., Dublin.
Mosley, Sir Oswald, Bart., D.C.L., F.L.S., G.S., &c., Rolleston Hall, Staffordshire.
Murchison, Sir R. Impey, G.C.St.S., F.R.S., L.S., G.S., Mem. Imp. Acad. Sc. Petersburg, Cor. R. Inst. France, &c.
Nash, Mr. William, Tunbridge Wells.
Neale, Thomas, Esq., Tunbridge Wells.
Neison, Francis G. P., Esq., F.L.S., Pall Mall.
Nicholson, Richard, Esq., Rochdale.
Noble, J. W., M.B., Danett's Hall, Leicester.
Northampton, Marquis of, D.C.L., Trust. Brit. Mus., President of the Royal Society, F.L.S., S.A., G.S., &c., 145, Piccadilly Terrace.
Norwood, Thomas, Esq., Hastings.
Ogilvie, George, M.D., Aberdeen.
Oldham, Thomas, Esq., M.A., F.G.S., M.R.I.A., &c., Professor of Geology, Trinity College, Dublin.
Ormerod, W., Esq., Rochdale.
Palmer, W. H., Esq., 24, Bedford Row, London.
Pascoe, F. P., Esq., M.R.C.S., Trewhiddle, near St. Austell.
Peach, C. W., Esq., M.R.C.G.S., Fowey, Cornwall.
Penneck, Rev. H., M.A., Penzance.
Percy, John, M.D., F.R.S., M.B.S.E., Birmingham.
Pim, James, Esq., Dublin.
Plomley, Francis, M.D., Ph.D., F.L.S., &c., Maidstone.
Pollexfen, Rev. J. H., B.A., M.D., 4, Belford Place, Clapham Rise.
Powell, Hugh, Esq., Seymour Place, Euston Square, New Road.
Price, John, Esq., Linen Hall Place, Chester.
Quekett, Edwin J., Esq., M.R.C.S., F.L.S., Lecturer on Botany, &c., 50, Wellclose Square.
Quekett, John Thomas, Esq., M.R.C.S., Royal College of Surgeons, London.
Rand, Mrs., Guildford.
Ranking, Robert, Esq., F.L.S., Hastings.
Ransome, R., Esq., Ipswich.
Reade, Rev. J. B., F.R.S., Stone Vicarage, Bucks.
Reed, Mr. William, Wadhurst.
Reeve, Lovell, Esq., F.L.S., King William Street, Strand.
Reeves, W. W., Esq., Tunbridge Wells.
Riley, John, Esq., M.B.S.E., Papplewick, near Nottingham.
Risk, A., Esq., Paisley.
Robertson, William, Esq., Quayside, Newcastle-on-Tyne.

Robinson, C. M., Esq., Stoke Newington.
Rogers, George, M.D., M.R.C.S., Bristol.
Rosling, Alfred, Esq., Newington Place, Kennington.
Ross, A., Esq., Featherstone Buildings, Holborn.
Rothwell, Miss, Adelaide Place, Ilfracombe.
Ruck, Mr. William, Baths, Cheltenham.
Russell, Frederick, Esq., Brislington, near Bristol.
Russell, Mr. J., Guildford.
Salter, Thomas Bell, M.D., F.L.S., Isle of Wight.
Salwey, Rev. T., M.A., Oswestry.
Sanders, William, Esq., F.G.S., Bristol.
Scott, Mr. Robert, Hill Park, Westerham, Kent.
Scouler, John, M.D., F.L.S., Professor of Geology in the Royal Dublin Society.
Shadbolt, George, Esq., Sutherland Street, Walworth.
Sharp, Mr. John, Tunbridge Wells.
Sidebotham, J., Esq., M.B.S.L., Manchester.
Simpson, S., Esq., F.B.S.E., The Greaves, Lancaster.
Skinner, Mr. Edward, Rye.
Smith, Mr. James, Coleman Street, City.
Smith, Rev. W., F.L.S., &c., Wareham.
Sollit, J. D., Esq., Grammar School, Hull.
Sopwith, T., Esq., F.R.S. & G.S., Allenheads, Northumberland.
Spence, William, Esq., F.R.S., L.S., &c., Lower Seymour Street, Portman Square.
Stackhouse, Miss Emily, Trehane, near Truro.
Staunton, Sir George Thomas, Bart., M.P., F.R.S., L.S., S.A. & H.S., Devonshire Street, Portland Place.
Stent, William, Esq., Fittleworth, Sussex.
Sterry, Charles, Esq., Trinity Square, London.
Stokes, C., Esq., F.R.S., L.S., S.A. & G.S., Verulam Buildings, Gray's Inn.
Streeten, Robert J. N., M.D., F.L.S., Worcester.
Stuart, William, Esq., 8, Frederick Street, Gray's Inn Road, London.
Symonds, F., Esq., M.R.C.S., Oxford.
Taylor, Richard, Esq., F.L.S., S.A. & G.S., M.R.A.S., &c., Red Lion Court, Fleet Street.
Tennant, James, Esq., F.G.S., &c., 149, Strand.
Thompson, W., Esq., President of Natural History Society of Belfast, &c.
Thwaites, G. H. K., Esq., M.B.S.L., Lecturer on Botany at the Bristol Medical School.
Topping, Mr. C. M., Pentonville, London.
Tosswill, Charles S., Esq., 8, Torrington Place, London.
Tovey, Mr., Blechynden Terrace, Southampton.
Tudor, R. A., Esq., M.R.C.S., Liverpool.
Turner, Col. L. H., Argyll Street.
Turner, Dawson, Esq., F.R.S., L.S., A.S., &c. Yarmouth.
Turner, Henry, Esq., Jeffery's Terrace, Camden Town.

Tyrrell, John, Esq., Totness.
Walcott, W. H. L., Esq., Clifton.
Wallis, Arthur, Esq., Brighton.
Walton, John, Esq., F.L.S., &c., 9, Barnsbury Square, Islington.
Ward, N. B., Esq., F.L.S., Z.S., M.B.S.E., &c., Wellclose Square, London.
Warington, Robert, Esq., Apothecaries' Hall, London.
Wells, W., Esq., Primrose House, Bootle, Liverpool.
West, Mr. Samuel, Tunbridge Wells.
West, W. James, Esq., M.R.C.S., F.G.S., M.B.S.L., &c., Tunbridge.
Westley, Mr., Regent Street.
White, Alfred, Esq., F.L.S., Cloudesley Square, Liverpool Road, London.
White, H. Hopley, Esq., Clapham.
Whitla, F., Esq., M.B.S.L., G.S.D., Belfast.
Wilkinson, M. A. Eason, M.D., M.B.S.E., Manchester.
Williams, Thomas, M.D., Swansea.
Windsor, John, Esq., F.L.S., Manchester.
Winton, F. C., Esq., Beckley, Sussex.
Woods, Joseph, Esq., F.L.S., F.S.A., G.S., &c., Lewes.
Woodward, Charles, Esq., F.R.S., President of Islington Literary and Scientific Society, 10, Compton Terrace, Islington.
Wright, Thomas, M.D., Cheltenham.
Yarrell, William, Esq., F.L.S., Z.S., &c., Ryder Street, St. James's.
Yates, James, Esq., M.A., F.R.S., L.S. & G.S., Upper Bedford Place, Russell Square.
Young, J. Forbes, M.D., F.Z.S., B.S., &c., Upper Kennington Lane, London.

ERRATA AND ADDENDA.

Page
61, to synonyms of *Desmidium Swartzii*, add
Desmidium bispinosum, Corda, *Observ. Micros. sur les Animalc. de Carlsbad*, p. 20. t. 4. f. 28 (1840).
74, for *M. morsa* read *M. americana* (Ehr.), and add the following synonym:
Euastrum americanum, Ehrenberg, *Verbreitung und Einfluss des Mikros. Lebens in Süd- und Nord-Amerika*, t. 4. f. 15 (1843).
Substitute *americana* for *morsa* twice in the description.
86, line 5, for δ. read β.
99, to synonyms of *Cosmarium Botrytis*, add
Euastrum interstitiale, Kützing, *Phycologia Germanica*, p. 136 (1845), according to Kützing *in lit*.
101, line 2, *dele* Sussex.
127, add, Sporangia of *Staurastrum hirsutum* have been gathered abundantly at West Point, New York, by Professor Bailey.
130, under *Staurastrum pungens*, *dele* the reference to the American Journal of Science and Arts.
213, under *Staurastrum pygmæum*, add, Mr. Jenner informs me that he has seen this species amongst other Desmidieæ which I had forwarded from Dolgelley; and Mr. Thwaites has given me its sporangia, gathered near Devonport by Mr. Dansey.

The following additions will render more complete the directions for mounting Algæ which have been given in the Introduction.

The rarer species of Desmidieæ are frequently scattered amongst decayed vegetable matter, so that it is difficult to procure good specimens for mounting. In such cases a small portion of the mass should be mixed with a little of the creosote fluid, and stirred briskly with a needle. After this has been done the Desmidieæ will sink to the bottom, when the refuse should be carefully removed. Successive portions having been thus treated, specimens will at length be procured sufficiently free from foreign matter. Even in ordinary circumstances, if a small extra quantity of fluid be placed in the cell, and the slide gently inclined, most of the dirt can be removed by a needle before the cell is closed, which process will materially increase the beauty of the preparation.

If the cells are insufficiently baked, the japan occasionally peels off the glass after the specimen has been mounted for some time. To obviate this inconvenience, Mr. Jenner previously heats the cell, with much caution, over a rushlight, until the japan becomes of a dark colour, and vapour ceases to arise from it.

When *gold-size* is used for closing the cell, the intrusion of some of it frequently destroys valuable specimens, whatever care may be taken: Mr. Jenner has therefore relinquished it, and now employs a varnish made of coarsely comminuted purified *shell-lac* or translucent sealing-wax, to which is added *rectified spirits of wine*, in sufficient quantity to cover it. This varnish will be ready for use in about twelve hours; when it is too thick a little more spirit should be added. Mr. Jenner applies three coats of this varnish, and about a week afterwards a fourth composed of *japan varnish* or *gold-size*.

I have tried this method extensively, and have never found my specimens spoiled by the varnish insinuating itself into the cell. This process requires less time, and herein it possesses another great advantage over the gold-size method, for the second coat being applied within half an hour, the risk of admitting air into the cell is much diminished.

To preserve the brush in a fit state it should always be cleaned with spirits of wine whenever it has been used.

At page 16 of the Introduction I have stated that I had never witnessed the circulation in the *Closteria*; but since the printing of that part Mr. Bowerbank has shown me the circulation in *Closterium Lunula* and in *Penium Digitus*. It seems, at least in the *Closterium*, to be restricted to the space between the mass of endochrome and the integument; for neither Mr. Bowerbank nor myself could detect it in the internal parts of the endochrome. I at first supposed that the circulation was confined to the margins; nor did I perceive it elsewhere until Mr. Bowerbank adjusted the microscope and showed me that the motion extended over the whole surface of the endochrome. The circulation being, as I have just stated, carried on between the integument and the mass of endochrome, which is usually brought fully into view by the observer, explains the difficulty experienced in detecting the circulation, except at the margins. The motion was very irregular: the fluid flowed at one time towards the extremities, and at another in the opposite direction, and the intervals between these changes were of uncertain duration. Streams also, though apparently not separated by any partitions, flowed side by side in contrary directions. The currents evidently consisted of an homogeneous fluid; but from time to time minute granules were detached from the internal mass of endochrome and carried along in the stream for short periods, after which they either returned to the quiescent portion or passed into other currents. It seems to deserve particular notice that the circulation was not interrupted at the suture. The process in *Penium Digitus* was somewhat different from that just described. The flow was continuous from the middle to the extremities, whence the returning streams apparently poured back through the centre of the mass of endochrome. I ought however to add that, as regards the *Penium*, I witnessed the movement only in one specimen, and in that but for a short period. I have subsequently succeeded now and then in obtaining with the triplet some obscure indications of the existence of a circulation; but for the exhibition of it with any accuracy or certainty the simple microscope seems altogether inadequate.

I have mentioned in the Introduction many points of similarity between Desmidieæ and acknowledged Algæ, and as every additional fact illustrating the resemblance tends to confirm the opinion advocated in this work, that the former belong to the vegetable kingdom, I gladly seize this opportunity of stating that Mr. Jenner has recently shown me in a species of *Tyndaridea* evident, though faint, longitudinal striæ, similar to those which are present in many *Closteria*, and that Mr. Bowerbank has pointed out to me the same appearance in a species of *Tiresias*.

ERRATA AND ADDENDA.

Professor Bailey has recently presented me with a collection of American Desmidieæ, preserved in Goadby's solution, in small sealed bottles; one of these, filled with specimens gathered in a pond near Princeton, New Jersey, is so rich in species, that I feel assured a list of them will be very acceptable to my readers, since it adds so many species to the great number which have been recorded in this work as common to both countries.

Hyalotheca dissiliens.
Didymoprium Grevillii.
—————— Borreri.
Desmidium Swartzii.
Aptogonum Baileyi.
Sphærozosma excavatum.
—————— pulchrum.
Micrasterias fimbriata.
—————— radiosa.
—————— rotata.
—————— truncata.
—————— Torreyi.
—————— furcata.
—————— muricata.
—————— oscitans.
—————— pinnatifida.
Euastrum oblongum.
—————— crassum.
—————— humerosum.
—————— Didelta.
—————— ansatum.
—————— rostratum.
—————— elegans.
—————— binale.
Cosmarium Cucumis.
—————— pyramidatum.
—————— Phaseolus.
—————— granatum.
—————— crenatum.
—————— ovale.
—————— margaritiferum.
—————— Brebissonii.
—————— tetraophthalmum.
—————— connatum.
—————— Cucurbita.

Cosmarium orbiculatum.
—————— moniliforme.
Xanthidium fasciculatum.
—————— cristatum.
—————— octocorne.
Arthrodesmus convergens.
Staurastrum muticum.
—————— dejectum.
—————— hirsutum.
—————— Arachne.
—————— gracile.
Penium margaritaceum γ.
—————— Digitus.
—————— interruptum.
—————— Brebissonii.
Docidium nodulosum.
—————— Ehrenbergii.
—————— Baculum.
—————— constrictum.
—————— verrucosum.
—————— nodosum.
—————— verticillatum.
Closterium Leibleinii.
—————— Dianæ.
—————— didymotocum.
—————— angustatum.
—————— costatum.
—————— striolatum.
—————— Ralfsii.
—————— setaceum.
—————— cuspidatum.
Pediastrum Heptactis.
—————— biradiatum.
—————— ellipticum.

INTRODUCTION.

INTRODUCTION.

Until a recent period, the study of the minute objects which form the subject of this work had been more neglected in this kingdom than almost any other branch of Natural History, and I commenced my researches with the intention of acquiring for myself some fuller and more satisfactory information in regard to disputed points in their history, and also with the hope that I might be able to present to the British Naturalist such a description of our species as seemed necessary towards making the knowledge of them at home keep pace with its advance on the Continent. I soon discovered not only that we possessed many species hitherto undescribed, but that various points in their economy, not devoid of interest, remained still unexplained or doubtful; and, rewarded beyond my expectations, I hastened to communicate the result of my investigations in a series of papers to the Botanical Society of Edinburgh. As these memoirs have received the gratifying approbation of distinguished naturalists, both in this country and abroad, I have been induced, at the solicitations of my friends, to publish as complete a monograph of the British species as the present state of our knowledge will permit.

The Desmidieæ have been for a long time a common territory, claimed both by zoologist and botanist. In consequence, a greater number of persons have devoted themselves to their study, and as they have often entered on the subject with different and conflicting views, every fact relating to their history has undergone a more rigorous examination.

The Desmidieæ are of an herbaceous green colour; a few only of the Closteria have the integument coloured, but in all the internal matter is green. All the family are inhabitants of fresh water. Mr. Thwaites indeed has gathered two or three species in water slightly brackish, but the same species are also found in localities remote from the sea. Certain marine objects that have been classed with the Desmidieæ have the internal matter of a brown colour, but these, in my opinion, belong to the Diatomaceæ.

Their most obvious peculiarities are the beauty and variety of their forms and their external markings and appendages; but their most distinctive character is the evident division into two valves or segments.

Each cell or joint in the Desmidieæ consists of two symmetrical valves or segments, and the suture or line of junction is in general well-marked; in a few instances however, as in Scenedesmus, it is determined principally by analogy. In Pediastrum its situation is shown by a more or less evident notch on the outer side, but no separation has been noticed. The structures of Scenedesmus and Pediastrum are in fact less known than those of the rest of the family; and of their modes of reproduction we are altogether ignorant. In the other genera the suture eventually opens and allows the escape of the contents, and it is indicated by either a transverse line or a pale band, and usually also by a constriction.

An uninterrupted gradation may be traced from species in which these characters are inconspicuous to those in which they are fully developed: thus in Closterium and some species of Penium there is no constriction; in Tetmemorus, in some Cosmaria, and in Hyalotheca, it is quite evident, although still but slight; in Didymoprium and Desmidium it is denoted by a notch at each angle; but in Sphærozosma, Micrasterias and some other genera, the constriction is very deep, and the connecting portion forms a mere chord between the segments, which appear like distinct cells, and are so considered by Ehrenberg and others.

When the papers on the Desmidieæ were publishing in the 'Annals of Natural History,' I stood alone in regarding the frond as a single cell, differing on this point not only from Ehrenberg, but from every author whose works I had seen. Professor Kutzing, however, in his 'Phycologia Germanica,' has by independent observations arrived at the same conclusion; an important corroboration of the correctness of the opinion I then advanced. That the frond in Euastrum and allied genera is really a single constricted cell, and not a binate one, will, I am persuaded, be apparent to any one who traces the gradations mentioned above; but as the opinions of such distinguished naturalists as Ehrenberg and Meneghini are deservedly of much influence, and the subject is so important,—since upon the view which we take of it depends the explanation of the division of the frond presently to be described,—I shall proceed to notice some facts which seem to me quite decisive.

In Navicula and other genera of Diatomaceæ the frustules are often truly binate, and, as each frustule is complete in itself, though they be separated from each other their respective contents will still be protected on all sides; and even should one be broken the contents of the other will be undisturbed. In the Desmidieæ, on the other hand, as there is no septum between the segments, if these separate, or an opening be made in one, the contents of both will escape; and I have more than once observed the minute granules passing from one segment through the connecting tube into the other. The conjugation of the fronds and formation of sporangia I believe to be altogether irreconcilable with the supposition of binate cells. For in the simple Algæ are many examples where the contents of two cells meet and form a compact seed-like mass; but I know of no instance in which the contents of more than two cells are thus united: nor does it appear probable that the process of reproduction in either an animal or a vegetable should require, or indeed permit, the conjunction of four individuals for that purpose; but if the fronds were binate, it must follow that the cooperation of four individuals would be necessary for the continuation of the species. In *Didymoprium Grevillii* the reproductive body is contained within one of the coupling joints; so that if they were binate we should have the contents of two cells passing into two other cells, the contents of all four uniting into one body, two of the cells at all events forming a single chamber. It may indeed be suggested that the joint in Didymoprium differs from the frond in some of the other genera; but such a supposition is utterly untenable, and will never be advanced by any one acquainted with these objects. In both the endochrome is divided into two by a pale transverse band marking the junction of the valves; and here they in both eventually open, and permit the escape of their contents. In the one case, as in the other, the coupling bodies alike communicate at this point, and the entire process is essentially the same. I will now presume that I have proved that the bipartite Desmidieæ are truly cells more or less constricted, and in the following details I shall so designate them.

In the Desmidieæ the multiplication of the cells by repeated transverse division is full of interest, both on account of the remarkable manner in which it takes place, and because it unfolds, as I believe, the nature of the process in other families, and furnishes a valuable addition to our knowledge of their structure and physiology.

The compressed and deeply constricted cells of Euastrum offer most favourable opportunities for ascertaining the manner of the division; for although the frond is really a single cell, yet this cell in all its stages appears like two, the segments being always distinct, even from the commencement. As the connecting portion is so small, and necessarily produces the new segments, which cannot arise from a broader base than its opening, these are at first very minute, though they rapidly increase in size. The segments are separated by the elongation of the connecting tube, which is converted into two roundish hyaline lobules. These lobules increase in size, acquire colour, and gradually put on the appearance of the old portions. Of course, as they increase the original segments are pushed farther asunder, and at length are disconnected, each taking with it a new segment to supply the place of that from which it has separated.

It is curious to trace the progressive development of the new portions. At first they are devoid of colour, and have much the appearance of condensed gelatine, but as they increase in size the internal fluid acquires a green tint, which is at first very faint, but soon becomes darker; at length it assumes a granular state. At the same time the new segments increase in size and obtain their normal figure; the covering in some species shows the presence of puncta or granules; and lastly, in Xanthidium and Staurastrum the spines and processes make their appearance, beginning as mere tubercles, and then lengthening until they attain their perfect form and size; but complete separation frequently occurs before the whole process is completed. This singular process is repeated again and again, so that the older segments are united successively, as it were, with many generations. In Sphærozosma the same changes take place, and are just as evident, but the cells continue linked together, and a filament is formed, which elongates more and more rapidly as the joints increase in number. This continued multiplication by division has its limits; the segments gradually enlarge whilst they divide, and at length the plant ceases to grow; the division of the cells is no longer repeated; the internal matter changes its appearance, increases in density, and contains starch-granules which soon become numerous; the reproductive granules are perfected, and the individual perishes. In a filament the two oldest segments are found at its opposite extremities; for so long as the joints divide they are necessarily separated further and further from each other. Whilst this process is in progress the filament in Sphærozosma consists of segments of all sizes; but after it has reached maturity there is little inequality between them, except in some of the last-formed segments, which are permanently smaller. The case is the same with those genera in which the separation of the cells is complete. I admit that the division of the cells just described apparently differs greatly from that in other simple Algæ; but I believe that the process in all is essentially the same, and that whatever differences exist are modifications necessarily resulting from the different forms of the cells. In the examples already given the cell itself consists of two distinct portions, having a constriction between them; hence each of the new-formed portions is similarly distinct from the older one which forms it and to which it is united.

In order fully to elucidate the subject, cells may be distributed into three principal kinds, distinguished by their form:

1st. Bipartite cells, already described, and more or less constricted at the middle;

2nd. Cells globose or rounded at the ends, or having the extremities attenuated;

3rd. Cylindrical cells.

Bipartite cells belong only to the Desmidieæ; cells globose or roundish at the ends are seen in the Nostocs and Palmelleæ; attenuated cells in the Desmidieæ; and cylindrical ones in the Conjugatæ, Tiresias, &c.

It is obvious that the new portions must arise from the whole of the junction margin of the original valves; consequently when the junction occupies only a part of the breadth the new portion will be narrower than the old; but when the junction of the valves is as broad as the cell, the new portion will from the beginning be of the same breadth. From this important fact, we may explain the different sorts of division. Since in the two latter kinds of cell the valves are united by their entire breadth, the new portions cannot be distinguished by their size, we must therefore have recourse to other aids to enable us to trace the changes and satisfy ourselves of their real identity with that already described; and I hope to be able to show that this identity does exist.

In Nostoc and Anabaina the cells are globular, and as there is no constriction we might remain ignorant of the real method of division; but, guided by the analogical process in the Desmidieæ, I hope to make it sufficiently plain. The hemispheres are thrust apart by the new formation; but now it is the outer rounded margin that we look to for an explanation. If a globe be cut into two equal portions, each will represent half a circle. By comparison with the neighbouring cells, we find that these two half circles remain unaltered, and are merely separated from each other, for if again brought together they would reconstitute the former globe. The new formations however separate them further and further, until the intervening space equals that occupied by the original globe, and then we find two globes exactly like the primary one, the internal half of each being the newly-formed one. During this time the inner portions, as they extend, develope more and more of the circle, until each becomes, as I have stated, a perfect hemisphere. The whole process cannot, of course, be seen in the same cell; but in a dividing filament some joints may be observed in one stage and some in another, which renders the evidence complete.

When the cell is oblong, or only rounded at the extremities, the process, though similar, is less evident: the cell at first seems merely to elongate until it obtains nearly twice its original length, when the division commences and the rounding of the new ends becomes apparent. The tapering cell presents but little difference, for the separation takes place before its extremities are fully developed. Sometimes these cells separate obliquely, as in Spirotænia and Scenedesmus.

I ought to state however that the opinions advocated above do not agree with those of M. De Brébisson, who has attained so high a reputation for his intimate acquaintance with the freshwater Algæ, and to whose kindness I have been so often indebted during the progress of the present work. He considers that there is an essential distinction in the mode of division between the Desmidieæ and Nostochineæ (including in the latter the Palmelleæ), and that from it indeed differential characters are obtained by which we can distinguish these nearly-allied groups. He observes of *Hormospora mutabilis*, Bréb.*, " Ils sont le plus souvent géminés, se multipliant par une division spontanée (déduplication) transversale, comme cela arrive dans quelques autres Pleurococcoidées. Une division analogue a lieu dans les Desmidiées, auxquelles on serait d'abord tenté de rapporter les Hormospora ; mais les demi-corpuscules (hémisomates) des Desmidiées développent à leur point de séparation une nouvelle portion semblable à la première, tandis que, dans l'accroissement des Nostocinées, les corpuscules sont divisés en deux par un étranglement transversal, sans qu'il s'ensuive une reproduction sur chacun des points de rupture. Il y a dans ce cas, comme je l'ai dit ailleurs, *déduplication simple*. Dans les Desmidiées, il y a *déduplication* et *réduplication*."

It is with unfeigned diffidence that I venture to dissent from the opinion of one possessing so profound a knowledge of these tribes, and I do so only from conviction, the result of close and repeated investigations.

I have stated my belief that the same changes occur in both the Desmidieæ and the Nostochineæ. A cell in Micrasterias has two hemispheres, just as a joint in Anabaina has ; in both these separate, and in both each hemisphere becomes again a perfect sphere ; and if in Micrasterias the two hemispheres were united by their whole bases, there would not remain even an apparent difference between them.

The form of the cylindrical cells no longer helps us in tracing the method of division. In Penium as in the Conjugatæ, they seem merely to elongate and then divide. As I formerly suggested, in a paper read before the Botanical Society of Edinburgh, I consider it extremely probable that in all the simple Algæ the cell or joint consists of two valves, and that additions occur at their junction, the original parts remaining unaffected : but this it may never be possible to

* Annales des Sciences Naturelles, Jan. 1844.

demonstrate satisfactorily, unless a species of Conferva with a coloured integument should be detected, or some means can be devised for permanently colouring the filaments without impairing their growth. Then indeed the question might be determined; at present I can merely show the probability that the cell in cylindrical species of Desmidieæ agrees with the joint in a Zygnema or Tyndaridea; since whenever the covering is colourless and free from markings not the slightest difference can be perceived. This is the case in a few species of Penium; and hence *Penium Brebissonii* is by some authors placed in the Palmelleæ. In *Penium margaritaceum* and *Penium Cylindrus* the integument is coloured, and we are enabled, by means of the paler appearance of the newly-formed portions, to satisfy ourselves that in these also each half of the original cell is acquiring during the division a new partner. In Didymoprium the same fact is rendered apparent, because the suture passes between minute teeth; these teeth recede from each other, and the new teeth which appear between them show the place where the separation of the joint has occurred.

The spontaneous division of the frond is included by some writers amongst the modes of reproduction; but this is not strictly correct, for it is rather the manner in which the individual plant grows, since all the cells arrive at maturity nearly at the same period and terminate their existence about the same time.

The Desmidieæ are most probably reproduced only in two modes; one by the escape of the granular contents of the mature frond, and the other by the formation of sporangia, the result of the coupling of the cells.

When the cells approach maturity, molecular movements may be at times noticed in their contents, precisely similar to what has been described by Agardh and others as occurring in the Confervæ. This movement has been aptly termed a swarming. It has been seen by numerous observers,—in this country by Messrs. Dalrymple, Jenner, Thwaites, Sidebotham, Dr. Dickie, and others. The cause of this sudden commotion cannot be ascertained; but I have met with it more frequently in specimens that have been kept some days than in fresh-gathered ones. When released by the opening of the suture, the granules still move, but more rapidly and to a greater distance. With the subsequent history of these granules I am altogether unacquainted, but I conclude that it is similar to what has been traced in other Algæ.

The second mode of reproduction is by coupling, and the formation of sporangia. A communication is established between two cells, and a seed-like mass is formed in the same manner as in the Conjugatæ. This is green and granular at first, but soon becomes of a homogeneous appearance and of a brown or even reddish colour. There are however some variations in the process in the two families which require notice. In the Conjugatæ, the cells conjugate whilst still forming parts of a filament; but in the Desmidieæ, the filamentous species almost invariably separate into single joints before their conjugation, and in most of the species the valves of the cells become detached after they are emptied of their contents.

In many genera the sporangia remain smooth and unaltered; in others they become granulated, tuberculated or spinous; the spines being either simple or forked at the apex. In fact a sporangium may pass successively through all these stages, and hence may so change its appearance that its different states are liable to be taken for sporangia belonging to different species. In Tiresias also we sometimes meet with sporangia bearing spines, but in that genus they are arranged like the spokes of a wheel, and not scattered as in the Desmidieæ. What is the nature of the sporangia, and why so complicated a process is necessary, since the species is also propagated by means of the granules or zoospores which escape from the ruptured cell, are questions to which we cannot, in the present state of science, return a satisfactory answer. The sporangia I consider capsules; and this view seems to be confirmed by the experience of

Mr. Jenner, who informs me that the covering of the sporangium swells, and a mucus is secreted, in which minute fronds appear and, by their increase, at length rupture the attenuated covering*. That some purpose distinct from that performed by the zoospores is served by the coupling of the cells and formation of the sporangium cannot be doubted: for, where we can trace the operations of nature, we find that nothing is useless or in vain; nor is it reasonable to suppose that this complicated process should fulfil no other purpose than one already provided for without it. The sporangia are most abundant in spring before the pools dry up; and I would suggest, as no improbable conjecture, that the zoospores may be gemmæ, analogous to those present in *Marchantia polymorpha* and *Lunularia vulgaris*, and that they possess merely a limited vitality, which is destroyed unless they are at once placed in circumstances favourable to their growth, whilst on the other hand, in the conjugated cells some important change takes place during the commingling of their contents and the formation of the sporangium, like what happens in the production of seeds in general, which renders the sporangia capable of retaining the vital principle uninjured throughout long periods of drought.

That the Closteria couple and produce sporangia, in a manner similar to the Conjugatæ, has been recorded by Turpin and other writers. Correct figures of some species in that state are given by Ehrenberg in his 'Infusoria,' and Meneghini mentions that Brébisson had detected it in Desmidium†, but I am not aware that the conjugation of other Desmidieæ was noticed by any writer before I published in the 'Annals of Natural History' full descriptions of the formation of sporangia in *Tetmemorus granulatus* and *Staurastrum mucronatum*. Subsequently examples have been detected in almost every genus, and we cannot hesitate to consider it charac-

* An example of this condition occurring in *Closterium acerosum* is figured in the Plate containing that plant.

† "Diatomatum more secedunt, hasque simul e latere copulari in speciebus nonnullis detexit cl. Brébisson."—Meneghini, Synop. Desmid. p. 203.

teristic of the family. In this country the conjugation of about forty species has been noticed by different observers, and M. De Brébisson informs me that he has gathered sporangia of the following species in France:—*Hyalotheca dissiliens, Didymoprium Grevillii, Staurastrum pygmæum, S. controversum, S. muticum, Cosmarium Botrytis, Closterium Lunula, C. setaceum, C. acutum,* and *C. lineatum.*

In defining the genera and species, I have made no use of the reproductive bodies, for as yet too little is known about them to render them available for that purpose, and in many cases we are still uncertain whether the mature form is yet known. They are likewise so early detached from the emptied cells, that it is often very difficult to determine to what species they belong*.

I have gathered sporangia of other species besides those mentioned in this work; but whenever they were not still adherent to the fronds I have thought it best to pass them over, lest I should be the cause of error.

That the orbicular spinous bodies so frequent in flint are fossil sporangia of Desmidieæ cannot, I think, be doubtful when they are compared with figures of recent ones. Indeed one celebrated geologist, Dr. G. Mantell, who, in his 'Medals of Creation,' without any misgiving had adopted Ehrenberg's ideas concerning them, has changed his opinion, and in his last work regards them as having been reproductive bodies, although he is still uncertain whether they are of vegetable origin.

Ehrenberg and his followers describe these bodies as fossil

* A principal use of generic and specific characters is to enable us to identify the species we meet with; and although a genus may be accurately defined by characters taken from the reproductive parts, yet if that definition can be tested only in a few rare instances, it will be inferior in real value to one which is derived from less important parts that are always present. My wish has been to render the present work a practical one, useful in the fields as well as in the study: I have therefore omitted the employment of such characters, and endeavoured to express those which I have adopted as concisely as may be compatible with usefulness.

species of Xanthidium, but no doubt erroneously, since their structure is very different. For the true Xanthidium has a compressed, bipartite, and bivalved cell, whilst these fossils have a globose and entire one.

The fossil forms vary like recent sporangia in being smooth, bristly, or furnished with spines, which in some are simple, and in others branched at the extremity. Sometimes too a membrane may be traced, even more distinctly than in recent specimens, either covering the spines or entangled with them.

Some writers describe the fossil forms as having been siliceous in their living state, but Mr. Williamson informs me that he possesses specimens which exhibit bent spines and torn margins, and thus wholly contradict the idea that they were siliceous before they were imbedded in the flint.

In the present state of our knowledge it would be premature to attempt identifying the fossil with recent species: it is better therefore, at least for the present, to retain the names bestowed on the former by those who have described them. A paper on fossil Xanthidia by Mr. H. H. White, containing descriptions of eleven supposed species, accompanied by characteristic figures, may be consulted with advantage[*].

In all the Desmidieæ, but especially in Closterium and Micrasterias, small, compact, seed-like bodies of a blackish colour are at times met with. Their situation is uncertain, and their number varies from one to four. In their immediate neighbourhood the endochrome is wanting, as if it had been required to form them, but in the rest of the frond it retains its usual colour and appearance. I cannot satisfy myself respecting the nature of these bodies, but I believe them either to arise from an unhealthy condition of the plant, or else to be parasitic.

The only account I have seen of the discovery of fossil

---

[*] Microscopic Journal, vol. ii. p. 35.

fronds of Desmidieæ is by Professor Bailey, who detected various species of Closterium and Euastrum in calcareous marls, collected in New Hampshire and New York by Professors Hubbard and Hall, and also in marl at Scotchtown, New York, by Mr. Connors*. Professor Bailey informs me, that the specimens from the last-named station, and in which he found several Closteria, Euastra, &c., were taken from below the bones of the *Mastodon giganteus*. As sporangia of the Desmidieæ and other membranous bodies in a fossil state have lately been detected by Mr. Deane and Dr. G. Mantell in the grey chalk of Folkestone, it is probable that a careful search in that neighbourhood would also bring to light the fossil fronds of the Desmidieæ.

The production of the Desmidieæ in newly-formed collections of water is involved in obscurity. The late Mr. Miller of Penzance pointed out to me an instance of this kind well-worthy of notice. He found *Hyalotheca dissiliens* and other species of this family in an old water-butt, which stood in a yard remote from any apparent station for the Desmidieæ, and derived its water from the clouds alone; and the question naturally arises, How came the Algæ there? The theory of spontaneous generation has never obtained currency in this kingdom, and for my own part I am not unwilling to acknowledge that there are mysteries in nature which we cannot penetrate. I can therefore only attempt to account for the appearance of the Desmidieæ under such circumstances in two ways,—by supposing either that the atmosphere contains countless myriads of the sporules of the Desmidieæ and other Cryptogamia, which vegetate only when they meet a congenial situation, or that the seeds are conveyed by means of aquatic insects, many of which, it is well known, roam during the night by means of their wings from one piece of water to another. The latter I consider the more probable conjecture.

The entire question of the vegetation of the conjugating

---

* American Journal of Science and Arts, vol. xlviii. p. 340.

Algæ is far from being understood. A few years back I paid considerable attention to the subject, but without arriving at any satisfactory conclusion. The *Staurocarpus cærulescens* is not uncommon near Penzance, is generally in large quantity where it occurs, and, from its peculiar colour, cannot escape detection; on these accounts I made it a principal subject of my observation. Although I have yearly gathered it in several pools, and the sporangia are always abundantly produced, I have particularly noticed during five or six years' observation, that it has never in a single instance reappeared in the same pool. At Dolgelley, where also in some years it is common, I met with the same result, with a single exception when I gathered it in one pool for two successive years. I have noticed the same fact with regard to *Zygnema curvatum*, and I believe it holds good in regard to most if not all the other Conjugatæ; but as they are more liable to be overlooked, I cannot speak of them with the same certainty as of the above-mentioned species. Algæ in running water commonly recur every season. I called Mr. Jenner's attention to the subject: we were alike unsuccessful in our attempts to ascertain the cause of this singularity. His observations in general agreed with mine, that the plant will not appear in the same pool for two successive years; but he found too many exceptions to justify any certain conclusion.

All the Desmidieæ are gelatinous. In some the mucus is condensed into a distinct and well-defined hyaline sheath or covering, as in *Didymoprium Grevillii* and *Staurastrum tumidum*; in others it is more attenuated, and the fact that it forms a covering, is discerned only by its preventing the contact of the coloured cells. In general its quantity is merely sufficient to hold the fronds together in a kind of filmy cloud, which is dispersed by the slightest touch. When they are left exposed by the evaporation of the water, this mucus becomes denser, and is apparently secreted in larger quantities to protect them from the effects of drought. I have observed more especially that *Tetmemorus granulatus* and *Penium Brebissonii*, under such circumstances, form a distinct mucous stratum; and on this account some authors have placed the latter with the Palmelleæ, although when it occurs in water it is less gelatinous than many other species belonging to the family of the Desmidieæ.

I have never obtained a clear view of the circulation witnessed by various authors in species of Closterium and Docidium, but I have no doubt that it is correctly described in the following account by Mr. Dalrymple of his own observations:—"A circulation of the fluids within the shell was observed independent of the active molecules; this was regular, passing in two opposite currents, one along the side of the shell, and the other along the periphery of the gelatinous body*." Professor Bailey observes, that "The account by Mr. Dalrymple agrees with what I have witnessed in several species. The currents are very distinct; so much so in fact, that they attracted my attention before I was aware that they had been noticed by others†."

Dr. Williams of Swansea has kindly translated from the German a paper on this subject by Labarzewski‡, but it is too long for insertion here, nor would it be intelligible without the accompanying figures. Labarzewski's observations were made on specimens of *Closterium Lunula*. The circulating fluid was clear and thick, and filled the space between the covering and the central mass of green granular material, from which granules detached themselves from time to time, and after moving along the margins, returned to their former situation. The current was quickened at the ends and near the suture, where it was lost, but reappeared in the other segment. The circulation was intermittent, lasting each time about seven seconds.

I now approach a question on which I feel the greatest anxiety, lest I should not do justice to the arguments of those from whose opinion I may differ, or should fail satis-

---

* Annals of Natural History, vol. v. p. 416.
† American Journal of Science and Arts, vol. xli. p. 300.
‡ Linnæa, 1840, p. 278.

factorily to impress upon my reader the reasons which have appeared to my own mind incontrovertible.

The question is,—are the Desmidieæ animals or vegetables?

The arguments I have seen advanced in support of their animal nature appear to me so inconclusive, when contrasted with those adduced in favour of their being vegetables, that, although in the course of a long scientific correspondence I have sought to become thoroughly acquainted with the facts relied upon by the advocates of the former opinion, I have at times almost doubted whether my distance from the metropolis may not have precluded me from the opportunity of hearing others of a more convincing description.

I will however claim the merit of being at least desirous of stating the case impartially, and I have in vain consulted some distinguished naturalists who differ from me, that I might learn whether their opinions were supported by other reasons than those which are so generally known, and to which I shall presently refer. I will also add, what may be a fact of some weight, that I formerly considered the Desmidieæ animals, and the Diatomaceæ vegetables, and that careful observations alone have in a great measure reversed my opinions. The Desmidieæ I now believe have as strong a claim as the Conjugatæ or Palmellæ can have to rank with the Algæ. On the other hand I consider the proper station of the Diatomaceæ very doubtful. They have at least as much right to a place in the animal as in the vegetable kingdom; and perhaps the safest course would be that adopted by several celebrated continental naturalists, who regard them as belonging to a distinct and intermediate group, and partaking almost equally of the characters of animal and of vegetable.

The chief reasons advanced by Ehrenberg and others for placing the Desmidieæ in the animal kingdom are the following:—that they exert a voluntary motion; that they increase by transverse self-division; and that the Closteria have at their extremities apertures and protruding organs continually in motion. Although two of these reasons apply only to the genus Closterium, I freely admit that if the Closteria can be proved animals, the question as to the other genera will be decided.

Few indeed hint at such a motive for adopting the opinion, nevertheless I feel persuaded that the animal-like forms of the Desmidieæ have had a great, though not avowed, influence on the determination of this question. Did we trust solely to the eye, we should conclude that objects so different in form and variable in appearance, were far more like animals than vegetables. Their symmetrical division into two segments; the beautiful disciform, finely-cut and toothed Micrasterias, the lobed Euastrum, the Cosmarium glittering as it were with gems, the Xanthidium armed with spines, the scimitar-shaped Closterium embellished with striæ, the Desmidium resembling a tape-worm, and the strangely insect-like Staurastrum sometimes furnished with arms, as if for the purpose of seizing its prey;—all these characters seem indeed to pertain more to the lower animals than to vegetables. We are thus induced, however unconsciously, to judge before examination, and we naturally search for arguments in support of our preconceived opinion instead of those which may elicit the truth. But experience has proved that form alone is a most deceptive guide, the implicit dependence on which has, in many similar instances, been the cause of error; and I believe that if a person unacquainted with what has been written respecting the two groups should look at the representations of the Desmidieæ, and examine the graceful and arborescent Zoophyte, having its branches to all appearance loaded with fruit that is periodically produced, matured, and shed, he would without hesitation place the Desmidieæ in the animal, and the Zoophyte in the vegetable kingdom. Nor could we wonder at such a decision, since in former times, even observers of high scientific attainments, judging by external appearances, did in fact class Zoophytes with the Algæ.

Again, Ellis, who first established the animal nature of

the Zoophytes, and thus dispelled one error, presently committed a similar fault by transferring along with the Zoophytes (simply on account of their agreement in certain external characters), Corallina and other genera which, it is almost unnecessary to add, have not until a recent period been restored to their proper place amongst the Algæ*. The above is a sufficient illustration of the danger of trusting solely to the evidence of sight: I shall endeavour to show that it would in the present case also lead to an incorrect conclusion.

That the Desmidieæ have been so long associated in the same family with the Diatomaceæ, whose proper position is so doubtful, presents another obstacle to the recognition of the claims of the former to rank with the Algæ. For when so many eminent observers,—botanists and zoologists, notwithstanding they differ widely in respect of the department of Natural History to which these forms should be assigned, —concur in classing them together, we naturally suppose that there must be valid reasons for such a course. I shall therefore point out the distinctions between these groups, and show that, at all events, they can no longer be united in one family, but must be separated, as they have been by Kutzing in his more recent work, as well as by others.

I have shown that the cell in the Desmidieæ consists of two valves united by a central suture, and that during its division the new-formed portions are interposed between these valves. The Desmidieæ are membranous, or should a few species contain silica, it is not present in sufficient quantity to interfere with their flexibility. They rarely have acute angles, and are seldom (if ever) rectangular. They are often deeply incised or lobed, warted or spinous. The internal matter is of a herbaceous-green colour, and starch vesicles abound in the mature cell. They couple and form either orbicular or quadrate seed-like bodies, and are remarkable for the resistance which they oppose to decomposition.

* See the introductions to Johnston's admirable works on the British Zoophytes and Corallines.

In all these respects they differ from the Diatomaceæ. In the latter each frustule consists of three pieces, one central and ring-like or continuous all round, and the others lateral. The division is completed by the formation of new portions within the enlarged central piece, which then falls off, or else by a new septum arising at the centre; but I believe that, in every case, the separation commences internally before it extends to the covering*. Their coverings, with very few exceptions, are siliceous, withstand the actions of fire and acids, and may be broken but not bent; the frustules are often rectangular in form, are never warted, and scarcely ever spinous. Their internal matter is usually brown when recent; and, although some species are greenish, or become green after they have been gathered, none are of a truly herbaceous colour. Their vesicles bear some resemblance to those in the Desmidieæ, but they are of a yellower colour, and no starch has been detected in them. The Diatomaceæ do not conjugate†, and in general they very soon give out an offensive odour.

I have preferred treating of these topics first and at some length, because I wish the reader to enter upon the subject unprejudiced, so that after a fair examination of the conflicting arguments he may be ready to surrender his judgement to that which is supported by facts of greater weight.

The first reason advanced by Ehrenberg in support of the animal nature of the Desmidieæ is, that they possess a "voluntary motion;" but I must protest against his use of the term *voluntary* as prejudging the matter in dispute and assuming more than in the present state of knowledge can be ascertained; more indeed than he has attempted to prove.

That the Desmidieæ move must be admitted; for this fact has been noticed by too many accurate observers to permit

* For detailed descriptions of the mode in which cells are multiplied by division in the Diatomaceæ, see the Transactions of the Botanical Society of Edinburgh, vol. ii. p. 37.

† Since the above was in type Mr. Thwaites has detected four species of Diatomaceæ in a conjugated state; the sporangia are elongated and in pairs, and the internal matter is similar to that of the frustules.

any doubt of its truth, and although I have myself failed to perceive their actual movement, I have sufficient evidence of its occurrence. But, whilst making this admission, I must still maintain that in the lower tribes of organic life motion is not an indubitable sign of an animal nature*, and that the movements of the Desmidieæ must be very sluggish, or exercised only under peculiar circumstances, since I have never witnessed it, notwithstanding I have almost daily living specimens under my inspection. Mr. Jenner has been equally unsuccessful, and several friends, experienced in the use of the microscope, either have not seen it or speak of it in uncertain terms†.

Professor Bailey states that " their power of locomotion

---

* " Motum non determinare limites regni animalis exinde patet, quod sunt animalia, quæ non moventur, vegetabilia in quibus motum vividum videmus." Agardh, Conspectus Diatomacearum, p. 4.

" The active motions in plants and their parts, especially in Algæ, ought not to give rise to the supposition of an animal nature, even when they are called infusorial or animal motions."—Ehrenberg, Taylor's Scient. Mem. vol. i. p. 566.

† I subjoin the opinions of various observers :—

" Actual motion, arising from internal causes, I saw only in Sphærastrum ; and the slight movement, supposed to have been observed in some of the genera, is certainly of the same description as that of some Confervæ."—Meyen, 1839, as quoted in Pritchard's Infusoria, p. 180.

" These are animals instead of plants, if the faculty of locomotion will entitle them to that rank."—Carmichael, in Hook. Br. Fl. vol. ii. p. 398.

" It was impossible to determine whether the vague motions of Closterium were voluntary or not."—Dalrymple, see Annals of Nat. Hist. vol. v. p. 417.

" Motions apparently voluntary are easily seen . . . . I have seen *Euastrum margaritiferum* move quite distinctly."—Bailey, American Bacillaria.

" Elles n'ont pas un mouvement sensible sur le porte objet du microscope. Cependant il est facile de remarquer dans les localités où elles vivent où dans les vases où on les conserve, qu'elles se dirigent vers la lumière et se rapprochent en pellicules où en sortes de pinceaux d'un beau vert, réunies entre elles au moyen d'un mucus qui les entoure ordinairement."—Brébisson, in Chevalier's ' Des microscopes et leur usage.'

" The various species of Closterium, as well as the closely allied Euastra, have a distinct motion which cannot be referred to any extrication of gas. I have had species of Closterium and Euastrum confined in a compressor, in water perfectly free from other bodies, and they moved so fast that I found it impossible to sketch their forms by the camera lucida until they were killed."—Professor Bailey in lit.

may be rendered apparent by taking a portion of mud covered with Closteria, placing it in water exposed to light, and then if the Closteria are buried in the mud they will soon work their way to the surface, covering it again with a green stratum*." I have myself frequently observed this fact, especially in *Tetmemorus granulatus* and *Penium Brebissonii*; but I presume that it is owing to the stimulus of light rather than to any voluntary effort; at any rate the same result will follow if an Oscillatoria be substituted for the Closterium. In like manner, in some species of Nostoc, Anabaina, and Palmella, the filaments or cells throng towards the light; and should the specimen be turned over, they will in a few hours appear on the new surface, whilst they have become less numerous on the previous upper one. Of course the gelatine must be in a sufficiently lax state to permit such a movement.

Another proof of their power of locomotion is afforded by their retiring, in some instances, beneath the surface when the pools dry up. I have taken advantage of this circumstance in order to obtain specimens less mingled with foreign matters than they would otherwise have been. If a species be much mixed with mud, I take a saucer, fill it with earth made into a paste with water, and cover it with a piece of linen; over this I spread a thick layer, containing the Desmidieæ, and allow it to become nearly dry; within a few days the specimens will form a stratum on the linen, and may be scraped off with a knife. This plan however proves successful only with the smaller species, and minute Algæ accompany the Desmidieæ which are so obtained.

It is thus evident that whatever be the motive power of the Desmidieæ they possess it only in common with acknowledged Algæ, and in a less degree than either the Diatomaceæ, the Oscillatorieæ, the sporules of various Algæ, or indeed their own sporules.

Ehrenberg considers that increase by voluntary division is the character which separates animals from vegetables[†], and,

---

* American Bacillaria, in American Journal of Science and Arts, vol. xli. p. 300.

† Annals of Natural History, vol. ii. p. 123.

in fact, he produces no other reason for denying the vegetable nature of some genera. But it will probably be thought a sufficient answer to mention that Meyen and others have proved the growth by elongation and bisection of the cells to be very frequent, if not universal, in the more simple Algæ*.

Meyen remarks that Ehrenberg has interpreted " all the facts known as if these creations were undoubtedly animals, whilst the same facts would indicate quite a different signification if we proceeded upon the supposition that they were nothing but plants†." In the present instance we have a striking example of the truth of this observation, since Ehrenberg, on observing that division occurs in the lower animals, too hastily draws the inference that only animals can be the subjects of it.

On the other hand, Mr. Shuttleworth, in a letter to Mrs. Griffiths, written in the spring of 1842, described the growth of the simple Algæ by repeated spontaneous division of their cells, and illustrated it by sketches of *Zygnema nitidum*, &c. He states that the process is most evident in the Oscillatoriæ, in *Conferva fugacissima*, in *Lyngbya muralis*, and in the Conjugatæ; and, after noticing the similar process in the Diatomaceæ, he adds, " I trust that now their vegetable nature is beyond all doubt;" thus showing that from the same premises two able observers came to exactly opposite conclusions. It is right to mention that Mr. Shuttleworth did not claim the credit of being the original discoverer, but merely said that he had convinced himself of the correctness of the facts published.

To Mr. Hassall is justly due the credit of first directing the attention of British naturalists to this mode of growth;

---

* "The increase by self-division occurs in all these genera; this process is looked upon by Ehrenberg as one of the strongest and most decisive characters of animal nature; but I have elsewhere proved, in the most satisfactory manner, that self-division is very common, both in the lowest plants as well as in the elementary organs of the more highly developed ones."—Meyen, 1839; see Pritchard's Infusoria, p. 180.

† See Pritchard's Infusoria, p. 178.

and, as it appears he was not aware that it had previously been noticed by continental botanists, another proof is thus afforded that it has attracted the attention of independent observers.

The fact of the transverse bisection of the Algæ is now so well established that this character is employed by J. G. Agardh, in his generic definitions*.

In a preceding part of this introduction, I have endeavoured to show that in the Desmidieæ the process is the same as in all the simple Algæ.

If some should think that I have dwelt too long upon this topic, I trust I shall stand excused when it is considered that the celebrated Ehrenberg places much reliance upon this character, and that great weight is justly allowed to the authority of one who has devoted so much time and skill to the examination of the Desmidieæ; moreover, that his arguments are familiar to the British naturalist, and his opinions extensively embraced, whilst the opinions of those who differ from him on this subject are comparatively unknown.

I am not in a position either to deny or to affirm with confidence the presence of openings in the extremities of the Closteria, for in objects so minute it is very difficult, frequently perhaps impossible, to distinguish between a depression and an opening†. It appears indeed to me that in Closterium there is a slight notch, or more usually the rudiment of one, at the apex of the segments, a mere indication in short of what is fully developed in Tetmemorus and Euastrum; for in *C. Dianæ*, and some other species, there is an evident though minute notch, and in Penium not even a trace of it can be detected. In no instance can any portion of the contents of the cell be forced out from the extremities.

---

* "*Tiresias*, fronde tota homogenea et articulis omnibus continua subdivisione iterum iterumque divisis, atque coniocystis inclusis insigne."—Ag. Icon. Alg. Ined.

† "No one accustomed to the use of the microscope can be otherwise than aware how much very minute objects seen under a high power are apt to assume a character in accordance with preconceived notions."—J. T. Smith, in Annals of Natural History, vol. xix. p. 2.

Mr. Dalrymple believes that orifices exist which are closed by internal membranes.

Ehrenberg, on the contrary, describes not merely open orifices, but protruding organs or feet immediately behind them, and thus affords an example of his want of caution when he gives the rein to his poetic fancy; for in this instance at least he is clearly in error. Mr. Dalrymple, whose remarks on the Closteria are the undoubted result of careful examination, carried on with the desire of recording actual facts rather than of supporting a theory, denies the existence of any papillæ or proboscides at this part, and also the supposition of Ehrenberg, that the moving molecules seen near the extremities constitute the basis of such papillæ. This admission is the more valuable as coming from a person who is not only an accurate observer but a believer in their animal nature. Professor Bailey also informs us that he has failed to detect them.

In most of the Closteria there is at each extremity of the endochrome a distinct globule containing moving granules. Ehrenberg seems to have mistaken these for organs of motion. In *Closterium rostratum* and *C. setaceum* these granules are situated at a distance from the extremities, and, as they are apparently not contained within a globule, move more freely and afford better opportunities for observing them than those in the other species do. In one instance I saw them continue to move after escaping from the frond. Did the Closteria possess mouths we should expect to find them in every species; but *Penium interruptum*, in which there are conspicuous terminal globules and moving granules, has decidedly nothing of the kind; and as its end view is circular and rather turgid no more desirable plant could be selected for examination. If mouths are present they ought to be visible in this and the allied species, but not a single punctum can be detected in the end view of the empty frond.

I have now passed in review the four points advanced by Ehrenberg in support of the animal nature of the Desmidieæ, and I submit that I have shown that from them no evidence approaching to a satisfactory proof of his position can be deduced. In offering these remarks I hope I shall not be suspected of the least wish to undervalue Ehrenberg's discoveries. I believe that he is sometimes too hasty in arriving at his conclusions, and that his knowledge of the Desmidieæ is less accurate than of the Diatomaceæ and other tribes which have come under his observation. When we consider the extent of his researches, it is not surprising, eminent as he is, that he should fall into some errors, which a person inferior in ability, but confining his whole attention to a comparatively smaller range, is likely to detect. Nor even in respect of this family, although I must consider it the most defective portion of his great work, are the obligations slight which we owe to Ehrenberg: for not only has he enlarged our knowledge of the Desmidieæ by the discovery of many new facts, by his discrimination of the species, and by producing better representations than those which previously existed, but he has given an impulse to the study, and rendered it popular.

The fifth volume of the 'Annals of Natural History*' contains an abstract of a paper by Mr. Dalrymple on the Closteria, read before the Microscopical Society of London, from which I quote the following reasons for placing them in the animal kingdom :—

" 1st. That while Closterium has a circulation of molecules greatly resembling that of plants, it has also a definite organ, unknown in the vegetable world, in which the active molecules appear to enjoy an independent motion, and the parietes of which appear capable of contracting upon its contents.

" 2nd. That the green gelatinous body is contained in a membranous envelope, which, while it is elastic, contracts also upon the action of certain reagents whose effects cannot be considered purely chemical.

" 3rd. The comparison of the supposed ova with cytoblasts

* P. 415.

and cells of plants precludes the possibility of our considering them as the latter, while the appearance of a vitelline nucleus, transparent but molecular fluid, a chorion or shell, determines them as animal ova. It was shown to be impossible that these eggs had been deposited in the empty shell by other infusoria, or that they were the produce of some entozoon.

" 4th. That while it was impossible to determine whether the vague motions of Closterium were voluntary or not, yet the idea the author had formed of a suctorial apparatus, forbad his classing them with plants."

I confess I am unable to refer to any example in other Algæ of terminal globules like those present in the Closteria, but neither can one be found amongst animals; and if in some respects they have an analogy with organs belonging to the latter, in others they agree better with vegetable life. The contained granules seem to me to differ in no respect, except in position and uninterrupted motion, from other granules in the same frond, and, as I have already stated, I once saw the motion continue after their escape from the cell precisely as in other zoospores. Meyen observes that " the functions of these bodies is very difficult to determine, but they are to be found in very many Confervæ, and are perhaps to be likened to the spermatic animalcules of plants."

The contraction of the internal membrane of the Closteria, or the expulsion of their contents on the application of iodine or other reagents, cannot be relied upon as a satisfactory test for determining their nature, for the blandest fluids will, in some cases, occasion violent action. If fresh water touches *Griffithsia setacea*, the joints burst and spirt out their contents; and if it be applied to a species of Elachistea, the granular contents are instantly thrown into commotion. In certain conditions of the Closteria themselves, water will produce effects like those attributed to reagents. I have frequently witnessed that by the addition of water to specimens of Closteria, which had for some time been kept merely in a damp state, the frond has been ruptured, and the endochrome has escaped with considerable force. This is especially the case with *Closterium Lunula*; and the same circumstance occurs in most of the Desmidieæ, although in none of the others so remarkably as in that genus. Mr. P. Grant has favoured me with an account of numerous experiments which he instituted to determine the effects of different substances on the Desmidieæ, Conjugatæ, &c. From these it seems that the action of a reagent cannot be predicated with any certainty, and that the molecular motion is not affected by several strong poisons, whilst it yields to other substances less generally deleterious.

With regard to the "supposed ova," I fully agree with Meyen that "they are similar to the green corpuscles found in the cells of the Confervæ," and, when both are removed from their cells, I am persuaded that it would be impossible to distinguish between them. That they contain amylum has been, with good reason, pointed out by Meyen as decisive against the notion that they are eggs.

The latest advocate of the animal nature of the Desmidieæ is C. Eckhard, in his memoir 'on the Organization of the Polygastric Infusoria,' published in Weigmann's Archiv, part 3, 1846. With this memoir I am acquainted only through the medium of a translation by Dr. J. W. Griffiths in the 'Annals of Natural History*.' It is apparently written for the purpose of confirming Ehrenberg's views, and accordingly defends his accuracy in every respect. So far as regards the Desmidieæ, Eckhard notices only the Closteria, and he advances nothing additional except an opinion that the transverse suture is a fissure or mouth. In order that both sides of the question may be compared, I extract his remarks, and merely point out that he has left unnoticed the facts on which botanists place their chief reliance:—

"The grounds for their being of animal nature are derived partly from their motion, partly from their organization. On the leaves of Ceratophyllum, I observed the manner in

* Vol. xviii. p. 433.

which several Closteria adhered elegantly by one extremity; in about a quarter or half an hour many of them were situated in the same manner upon a higher part of the leaf: not a single animalcule was found on the side of the leaf, nor adherent longitudinally to it. They had evidently moved during the above time from the lower to the upper part of the leaf. If we observe their motions under the microscope they are not so rapid as those of many other polygastric infusoria, but the motion is always evidently animal. They swim, especially in summer, in the *most varied* directions, and I have frequently seen *Cl. acerosum* and *Lunula* swim *against* the current when the water on the object-holder was flowing towards one side, whilst fragments of plants, various kinds of Spirogyra and Oscillatoria were carried away. It is difficult here to discover anything but animal motion; to explain this however by electricity, as Turpin attempted, is unnatural, and not less absurd than that of the muscular fibre by the same natural agent by Strauss. But the relations of the organization of the Closterina are likewise in favour of their animal nature. In illustration of this I shall confine myself to *Cl. acerosum*. We see that the animal, which is expanded in the middle, is elongated symmetrically on each side. In the middle there is a transverse fissure, which probably serves for the admission of nourishment; since, when this animal is kept for some time in coloured water, we perceive little accumulations of the colouring matters. At the extremities we see on each side a vesicle, in which minute granules (?) incessantly move. In other species there is moreover a small aperture; it is situated more posteriorly, and is perhaps connected with the cell. Ehrenberg twice saw in this animalcule filaments (feet?) project from it. Internally there are on each side two or four cords, and a row (in other kinds several) of granular bodies. In the species figured, I have so often seen the above change in relative position, that I have been compelled to wait until they again appeared in their original position in order to delineate them. All this is not plant-like; and if the carapace of the Closterina should prove to be of a horny nature, as would appear to be the case from their becoming wrinkled when heated, they would be removed from the vegetable kingdom with still greater certainty."

Although at first sight it seems to indicate the contrary, the swarming of the zoospores or granules really affords a strong confirmation of the vegetable nature of the Desmidieæ. This swarming has been noticed in the Algæ by many eminent observers, and notwithstanding the extraordinary phænomena which it presents, no fact in their history is more firmly established. Hence J. G. Agardh, who so worthily emulates his father's fame, has been induced to apply the name of "Zoospermeæ" to one of his three primary divisions of the Algæ. In this country the swarming has been witnessed by Mr. Borrer, the Rev. M. J. Berkeley, Mr. Hassall, and others. I have frequently seen it in *Sphæroplea crispa* and *Draparnaldia tenuis*. A commotion arises within the cell as if all its contents were suddenly endued with life; the disturbance increasing, the cell opens, when the zoospores hasten from their prison, and, apparently enjoying their newly-acquired liberty, dart about in every direction, until, tired of their sport, they at length resume a quiescent state*.

This description has such a marvellous character that the reader may suppose it is somewhat indebted to the imagination, and I believe no one can witness the occurrence for the first time without being startled and almost led to doubt the evidence of sight. Such movements are so contrary to our ordinary experience of vegetable life, that we involuntarily hesitate to admit their compatibility with it; and on the continent many eminent naturalists, unable to find a satisfactory explanation, consider that in this stage the zoospores are really animals, and do not acquire their vegetable nature until a subsequent period. This opinion never obtained countenance in this country: yet a Berkeley did not

---

* The fullest details accessible to the English reader will be found in Mr. Harvey's excellent introduction to his 'Manual of British Algæ,' and in the introduction to Hassall's 'British Freshwater Algæ.'

esteem its refutation unworthy of his pen, and a Harvey thought it necessary to record his dissent from it and his belief that the phænomenon must be regarded as a "strictly vegetable peculiarity." All that I have alluded to as happening in the Algæ occurs also in the Desmidieæ; but no similar motion has ever been observed in the contents of an animal after their escape from the individual, and I therefore claim its presence in the Desmidieæ as a strong presumptive evidence in support of their vegetable nature.

Starch has been observed in the Desmidieæ by many persons, and I am not aware that its occurrence is now denied by any one. I should therefore have supposed that it hardly required further proof, but as the subject is altogether unnoticed by most writers on the opposite side, it becomes necessary to bring it fully before the reader.

Meyen first directed attention to the presence of starch as a conclusive proof that the Desmidieæ are Algæ. He states that in several genera he has "distinctly seen that the large and small granules contained amylum, and were sometimes even entirely composed of it[*]," and that "in the month of May he had observed many specimens of Closterium in which the whole interior substance was granulated, and all the grains gave with iodine a beautiful blue colour, as is the case with starch, *which is not an animal product.*" I know not that Ehrenberg or his followers have ever repeated Meyen's experiments or taken the slightest notice of them, important as they are to the solution of the disputed question. In this country, Mr. Dalrymple at first failed to detect starch in the Closteriæ, but afterwards observed it in specimens of *Closterium* (*Penium*) *Digitus*, which I sent him, and acknowledged its presence in the following terms :—" I have examined the specimens sent up, and in several I can detect the blue colour of the iodide of starch; this is by no means however universal, some being merely stained yellowish-brown; but in those instances there appears to be an absence

---

[*] See American Journal of Science and Arts, vol. xli. p. 298.

of granular matter, the fact of blue granules in some is however decisive of the presence of starch." Mr. Dalrymple is so careful and accurate an observer, that I doubt not his failure was caused by the immature state of the specimens on which he experimented, as he indeed subsequently allowed might have been the case.

Professor Bailey of the U.S. Military Academy, in an article on the American 'Desmidiaceæ'*, gives copious extracts from Mr. Dalrymple's paper on the Closteria, accompanied by his own remarks. He bears testimony to the general correctness of Mr. Dalrymple's observations, but with regard to those on the action of iodine, he says, "I cannot otherwise account for Mr. Dalrymple's statement, that iodine 'in no instance produced in the Closteria the violet or blue colour indicating starch,' than by supposing that the specimens he examined were not in the proper state to exhibit it. Meyen expressly states that it is 'at certain times, particularly in spring,' that the starch may be detected. I am able by conclusive experiments to *confirm* Meyen's statement as to the presence of starch in these bodies. In specimens gathered in November I find no difficulty in producing the blue colour with tincture of iodine. Sometimes however the specimen becomes so opake by the action of this reagent, that the purple colour of the granules can only be detected after crushing the specimen by means of the compressor. *The characteristic colour of iodide of starch is then shown most distinctly.* I have repeatedly treated in this way *Closterium Trabecula*, as well as others, and have uniformly found that a portion of the interior takes the purplish colour†."

I have myself repeatedly noted the effects of iodine on many of the Desmidieæ. In a young state the cells are filled with a green homogeneous fluid, which as the plant approaches to maturity becomes granular. Scattered amongst

* American Journal of Science and Arts, vol. xli. p. 287.
† Ibid, p. 301.

this minutely granular matter, larger granules, or rather vesicles, make their appearance. These Ehrenberg calls ova; but I cannot perceive the slightest difference between them and the granules present in the higher Algæ.

On applying diluted tincture of iodine to different species of the Desmidieæ, the granules become very dark with a purplish tinge showing the presence of starch. When the tincture of iodine is used in its undiluted state, the colouring matter becomes nearly black and conceals the bluish tint; in some specimens too this tint is hardly perceptible, whilst in others it is very apparent. In no instance have I found the presence of starch indicated unless granules were present, as the fluid colouring-matter always becomes brownish. The application of iodine to Conjugatæ in different stages of growth was followed by a precisely similar result. In the young plant no starch was detected, but the colouring matter changed to an orange-brown. On the other hand, in the conjugating filaments the granules became blue, and the sporangia especially acquired the very dark colour observed in the Desmidieæ, and did not exhibit the iodide of starch until they were crushed.

Professor Bailey's testimony is the more satisfactory because it is not the evidence of a partisan, but the admission of one inclined to regard the Desmidieæ as pertaining to the animal kingdom. Although he does not "consider the presence of starch in these bodies as conclusive evidence that they are plants," yet as a professor of chemistry he is aware that starch is not an animal product, and he suggests therefore, with some ingenuity, "Is it not possible that they are animals which feed, wholly or in part, on amylaceous matter extracted from the aquatic plants among which they live? if so, the detection of starch in their stomachs is not surprising."

In the young cell there is no starch, but after its first appearance it continues to increase and is most plentiful in the sporangium; facts quite incompatible with Professor Bailey's suggestion, but strictly analogous with its presence in the Conjugatæ, and indeed in plants in general, for it is well known that starch is often most abundantly produced in the seed.

Of all the facts which indicate the vegetable nature of the Desmidieæ, this is undoubtedly the most important, since it is the most easily subjected to the test of experiment. The swarming of the zoospores is seen only now and then, and it is not always easy to trace the process of coupling; but every person can apply the test for starch, and needs only to bear in mind that unless granular matter be present there is no starch*.

The conjugation of the fronds in this family supplies an equally striking fact in proof that it belongs to the vegetable kingdom. I have already mentioned that we have amongst the Algæ many examples of the junction of two individuals†, and the commingling and union of their entire contents into a seed-like body for the purpose of reproduction. The case of the Conjugatæ is well known, and Mr. Hassall has proved that the same phænomenon takes place in other genera.

---

* After the diluted tincture of iodine is applied, the free iodine should be removed by the aid of heat, occasionally adding a little water to facilitate its removal. This in great measure removes the brownish stain which obscures the purple tint, and then on applying the highest powers of the microscope, the peculiar colour of the iodide can in general be easily perceived.

† So far as relates to the formation of the sporangium, each joint in the Conjugatæ must be considered a distinct individual. On the contrary, in the Desmidieæ, each frond resulting from division is merely a portion taken from the original one, and like a graft or slip from an apple, a willow, or a carnation, retains every accidental peculiarity of the variety from which it parted. Thus one pool may abound with individuals of *Staurastrum dejectum* or *Arthrodesmus incus*, having the mucro curved outwards; in a neighbouring pool every specimen may have it curved inwards; and in another it may be straight. The cause of the similarity in each pool, no doubt is, that its plants are offsets from a few primary fronds. The above fact, I must further remark, often renders it very difficult to decide whether a particular form be really a distinct species or merely a variety, since its abundance will not suffice to establish its claim to the rank of a species.

In the Desmidieæ, in like manner, a bag or cell forms between two individuals, and *the entire contents of both these* (or *indeed of four*, if we regard the fronds as binate) pass out and unite together to form one reproductive body, which becoming detached, leaves the parent corpuscles altogether empty. Such an occurrence is, I believe, not only unknown amongst animals, but is contrary to all our notions of animal propagation.

Among the Conjugatæ, Tyndaridea has an orbicular sporangium, Staurocarpus a cruciform or quadrate one; in Zygnema it is formed within one of the coupling cells, and in Mougeotia it is situated in the connecting tube. The Desmidieæ present us with corresponding variations. Their sporangia are generally orbicular, but Staurastrum, Tetmemorus, Closterium and Penium afford examples of cruciform and quadrate ones: and although the reproductive body is usually contained in the connecting tube, yet in *Didymoprium Grevillii* it is placed in one of the conjugating cells. Lastly, we have the *Didymoprium Borreri*, in which the conjugated filaments form a kind of network in the same manner as in *Mougeotia*.

That the Desmidieæ resist decomposition, exhale oxygen on exposure to the sun, preserve the purity of the water containing them, and when burnt do not emit the peculiar odour usually so characteristic of animal combustion, are other facts respecting this family, which taken singly might have less value, but in their combination furnish a most important support to the arguments already adduced.

On the foregoing statements I rest the claim of the Desmidieæ to be considered Algæ; and I confidently appeal to any impartial person whether they do not at least require to be answered before the conclusion which I have drawn can be rejected. But on consulting the works of those who profess to prove the contrary, we find that the important facts which I have here investigated are either altogether overlooked, or passed by without an attempt to controvert them. For my own part I believe them to be unanswerable.

I have pointed out that the swarming of the granules

occurs in acknowledged Algæ as well as in the Desmidieæ; and that nothing which can fairly be compared with it occurs in the animal kingdom. Can either of these assertions be disproved? Again, it has been seen that starch is abundantly produced in this family. Can a single example be referred to where it is an animal product? I have shown that the reproductive body is formed in a manner well known amongst the Algæ, but never detected in animals. Until these facts have been denied, or the arguments deduced from them refuted, I shall presume that the claim of the Desmidieæ to be considered vegetables is firmly established*.

The Desmidieæ I regard then as Algæ, allied on the one side to the Conjugatæ by similarity of reproduction, and on the other to the Palmelleæ, by the usually complete transverse division and by the presence of gelatine. Indeed the relation to the latter is so intimate, that it is difficult to say to which

* Since the above was written, a friend has furnished me with the translation of a passage from a recent work of Meneghini, which is so appropriate in support of my views that I gladly introduce it here:—

"The *Closteria* and the *Desmidieæ* are universally plants and not animals. In the actual state of science we are obliged to admit this proposition. The organic structure, the physiological phænomena, the history of their development, their chemical composition, manifest in these beings a perfect correspondence with others, which, under all their aspects, are comprised in the abstract idea of a plant. On the other hand, what they present in common with those beings evidently animal is but an appearance, or at most a similarity of external form. Ehrenberg was deceived by this appearance, and, guided by this fallacious resemblance, thought he discovered even in the Desmidieæ the same organic peculiarities which prove the animality of other beings. What must we infer from this? That even the most accurate observer and man of genius may err. This can never diminish his merit or render less important the services which he has rendered to science. The loss will fall only on those who, averse to the fatigue of observing, content themselves with the authority of the master, and embrace without distinction his real discoveries and his errors. Thank Heaven, the epoch of authority is passed, and whoever submits to its yoke may be allowed to err, because science will not advance the less for him, and may even derive advantage from those very errors. From the study of the Desmidieæ, and from their being brought into comparison with animals, valuable notions upon the intimate structure of vegetables have already been derived."—*Sulla Animalita della Diatomee*, p. 172.

family some genera belong. Thus Merismopædia, placed by Meneghini and Kützing in the Desmidieæ, I believe, with Meyen, to belong properly to the Palmelleæ. On the other hand, Cylindrocystis, placed in the latter by Meneghini and Brebisson, has been rightly removed to the Desmidieæ by Kützing and Hassall. Brébisson's discovery of conjugated specimens of *Coccochloris protuberans* and *Cocc. rubescens* brings to our notice another link connecting these families, whilst some species of Scenedesmus may be allowed to have an almost equal claim to rank with either.

Respecting the uses of the Desmidieæ little is known. Doubtless, in common with other aquatic vegetables, they tend to preserve the purity of the water in which they live, and Mr. Williamson has ascertained that to a great extent they furnish food for the bivalved mollusks which inhabit fresh waters.

As the Desmidieæ are unattached and very minute, they are rarely gathered in streams: nevertheless interesting species may occasionally be obtained where the current is so sluggish as to permit the thin retaining mucus to elude its force. In small shallow pools that do not dry up in summer they are most abundant; hence pools in boggy places are generally productive.

The Desmidieæ prefer an open country. They abound on moors and in exposed places, but are rarely found in shady woods or in deep ditches. To search for them in turbid waters is useless: such situations are the haunts of animals, not the habitats of the Desmidieæ, and the waters in which the latter are present are always clear to the very bottom.

M. de Brébisson informs me that in France, calcareous districts, which are so favourable to the Diatomaceæ, are very unproductive of Desmidieæ. I have myself had no opportunity of ascertaining whether the same fact obtains in England. Mr. Thwaites and Mr. Jenner have indeed furnished me with fewer habitats from calcareous soils, but the less frequent occurrence of small pools and bogs in such districts may perhaps partly explain the deficiency.

In the water the filamentous species resemble the Zygnemata, but their green colour is generally paler and more opake. They often occur in considerable quantity, and, notwithstanding their fragility, can generally be removed by the hand in the usual manner. When they are much diffused in the water, I take a piece of linen about the size of a pocket-handkerchief, lay it on the ground in the form of a bag, and then, by the aid of a tin box, scoop up the water and strain it through the bag, repeating the process as often as may be required. The larger species of Euastrum, Micrasterias, Closterium, &c., are generally situated at the bottom of the pool, either spread out as a thin gelatinous stratum, or collected into finger-like tufts. If the finger be gently passed beneath them they will rise to the surface in little masses, and with care may be removed and strained through the linen as above described. At first nothing appears on the linen except a mere stain or a little dirt; but by repeated fillings-up and strainings a considerable quantity will be obtained. If not very gelatinous, the water passes freely through the linen, from which the specimen can be scraped with a knife and transferred to a smaller piece; but in many species the fluid at length does not admit of being strained off without the employment of such force as would cause the fronds also to pass through, and in this case it should be poured into bottles until they are quite full. But many species of Staurastrum, Pediastrum, &c. usually form a greenish or dirty cloud upon the stems and leaves of the filiform aquatic plants, and to collect them requires more care than is necessary in the former instances. In this state the slightest touch will break up the whole mass and disperse it through the water. I would recommend the following method as the best-adapted for securing them. Let the hand be passed very gently into the water and beneath the cloud, the palm upwards and the fingers apart, so that the leaves or stem of the invested plant may lie between them and as near the palm as possible; then close the fingers, and keeping the hand in the same position, but concave, draw it cautiously towards the surface, when, if the plant has been allowed to slip easily and with an equable movement through the fingers, the Desmidieæ, in this way brushed off, will be found lying in the palm. The greatest difficulty is in withdrawing the hand from the surface of the water, and probably but little will be retained at first; practice, however, will soon render the operation easy and successful. The contents of the hand should be transferred at once either to a bottle or, in case much water has been taken up, into the box, which must be close at hand, and when this is full it can be emptied on the linen as before. But in this case the linen should be pressed gently and a portion only of the water expelled, the remainder being poured into the bottle, and the process repeated as often as necessary.

Sporangia are collected more frequently by the last than the preceding methods. When carried home, the bottles will apparently contain only foul water, but if it remain undisturbed for a few hours, the Desmidieæ will sink to the bottom, and most of the water may then be poured off. If a little fresh water be added occasionally to replace what has been drawn off, and the bottle be exposed to the light of the sun, the Desmidieæ will remain unaltered for a long time. I have now before me some specimens of *Euastrum insigne*, the fronds of which are in as good condition as when I gathered them at Dolgelly five months ago.

Mr. Thwaites's kindness has enabled me to render this Introduction more complete by the following account of two methods adopted by him in mounting minute Algæ for the microscope, which he has drawn up at my request. The remarks which I have appended have been derived from other sources, as well as from my own experience:—

"In making preparations of the Algæ for the microscope, there are two things which principally require to be attended to: first, to obtain a fluid which shall preserve the plant as little altered as possible from its appearance when living; and secondly, to adopt the best means for preventing the escape of this fluid after the object has been mounted in it.

With respect to the first point, the fluid which I have found to answer best is made in the following way :—To sixteen parts of *distilled water* add one part of *rectified spirits of wine* and a few drops of *creosote* sufficient to saturate it; stir in a small quantity of *prepared chalk* and then filter: with this fluid mix an equal measure of *camphor-water* (water saturated with camphor), and before using strain off through a piece of fine linen.

"This fluid I do not find to alter the appearance of the endochrome of Algæ more than *distilled water alone* does after some time, and there is certainly less probability of confervoid filaments making their appearance in the preparations; and there would seem to be nothing to prevent such a growth from taking place when the object is mounted in water only, provided a germ of one of these minute plants happen to be present, as well as a small quantity of free carbonic acid.

"Fluids containing a larger quantity of spirits of wine, and consequently of creosote also, than the one of which I have given the formula, produce a greater change in the appearance of the endochrome. I at one time thought that by increasing the density of the mounting fluid, the endochrome of the plant might be less disturbed, and I dissolved a small quantity of sugar in the fluid; but this made the cell-membrane too transparent, and rendered completely invisible the gelatinous sheath with which many species of Algæ are furnished.

" I now proceed to describe my method of making cells in which to mount preparations of Algæ. Some objects require very shallow and others somewhat deeper cells. The former may be made with a mixture of *japanners' gold-size* and *litharge*, to which (if a dark colour is preferred) a small quantity of lamp-black can be added. These materials should be rubbed up together with a painter's muller, and the mixture laid on the slips of glass with a camel-hair pencil as expeditiously as possible, since it quickly becomes hard; so that it is expedient to make but a small quantity at a time. For the deeper cells *marine-glue* answers extremely well, provided it is not too soft. It must be melted and dropped upon the slip of glass, then flattened, whilst warm, with a piece of wet glass, and what is superfluous cut away with a knife, so as to leave only the walls of the cell; these, if they have become loosened, may be made firm again by warming the under surface of the slip of glass. The surface of the cells must be made quite flat, which can be easily done by rubbing them upon a wet piece of smooth marble covered with the finest emery powder.

" When about to mount a preparation, a very thin layer of gold-size must be put upon the wall of the cell as well as on the edge of the piece of thin glass which is to cover it; before this is quite dry, the fluid with the object is to be put into the cell, and the cover of thin glass slowly laid upon it, beginning at one end: gentle pressure must then be used to squeeze out the superfluous fluid, and, after carefully wiping the slide dry, a thin coat of gold-size should be applied round the edge of the cell, and a second coat so soon as the first is dry: a thin coat or two of black sealing-wax varnish may then be put on with advantage, in order to prevent effectually the admission of air into the cell or the escape of fluid out of it.

" I at first mounted objects for the microscope without enclosing them in a cell previously prepared for their reception, but merely by laying them on the slip of glass with a drop of the fluid, and then covering them up with a piece of thin glass or talc, and afterwards surrounding the latter with a border of thick gold-size, in order to prevent the evaporation of the enclosed fluid. Preparations so made will frequently last some considerable time; but eventually the contraction, as it becomes dry, of the outer surface of gold-size forces the remainder, which still continues soft, between the two glasses, and the mounted object is thus injured. I found the same thing frequently to occur when the cells were made of gold-size only without the litharge; but this inconvenience seems to be completely obviated by the plan I have recommended above.

"I would remark that the gold-size employed should be of the consistence of treacle; when purchased it is usually too fluid, and should be exposed for some time in an open vessel, a process which renders it fit for use."

Mr. Sidebotham also has favoured me with directions for mounting the Desmidieæ. His method is nearly similar to that employed by Mr. Thwaites, but when the last coat of varnish is nearly dry, he applies a fine bronze with a camel-hair pencil. This not only has a neat effect, but prevents the risk of adhesion consequent on the softening of the varnish in warm weather. Mr. Sidebotham uses distilled water as the mounting fluid, but although his specimens retain the green colour of the endochrome better than any other I have seen, yet, for the reason assigned by Mr. Thwaites, I doubt whether such a mode is suitable for their permanent preservation.

Mr. Thwaites' fluid is superior to camphorated-water and various other liquids which I have tried; but as it requires more time for its preparation than may be at all times convenient, I find the following the best substitute:—

Bay-salt and *alum*, one grain each, dissolved in an ounce of *distilled water*.

Goadby's solution acts too powerfully on freshwater Algæ, and *corrosive sublimate* injures the specimens.

In mounting the Desmidieæ great attention is necessary to exclude air-bubbles, which cannot be avoided unless the fluid completely fills the cell; and also not to use too much fluid, as in this case the smaller species will often be washed away on the escape of the superfluous portion.

As the cells cannot be sealed whilst any moisture remains on their edge, it should be removed by blotting-paper, which is preferable to any other mode.

The thin glass manufactured for the purpose is preferable to talc for covering the specimen, and should always be used by those who possess an achromatic microscope; but with a simple instrument the triplet can be used only when talc is the covering.

Specimens are frequently spoiled by the intrusion of the gold-size into the cell. Experience will best teach how this accident is to be avoided; attention however to the following particulars will assist the inexperienced. The cells should be prepared some days before they are used, in order that their walls may become firm. When the cell is closed the brush should be passed round the edge of the cover, with just sufficient size to prevent the admission of air into the cell; and upon the operator's care in this respect will depend his success. If too little size be used the air finds admission, and if much be put on, or if the cell be not completely filled with fluid, the size will enter and spoil the specimen. When the first layer of size is quite dry, he should proceed as directed by Mr. Thwaites.

Mr. Topping has kindly sent me a description of his method of preparing cells for mounting microscopic objects. He uses strips of plate-glass of an uniform size (three inches by one), and marks on them the size of the cell, by taking two thin pieces of mahogany of the size of the glass, each having a hole (circular, oval or square, as may be desired) cut in its centre, the smaller corresponding with the inner margin of the cell, and the larger with the outer. These, when laid over the glass, offer a ready means of tracing with a diamond the space around the cell, which must then be filled up with japan. This is next hardened by placing the glasses in an oven, the heat of which should be raised gradually, as otherwise the japan will blister; but if care be taken in this part of the process the cells will resist the action of proof spirit.

The fluid which Mr. Topping has used for mounting consists of one ounce of rectified spirit to five ounces of distilled water, which he thinks superior to any other combination. To preserve delicate colours however, he prefers to use a solution of acetate of alumina—one ounce of the acetate to four ounces of distilled water. Of other solutions, he says that they "tend to destroy the colouring-matter of delicate objects, and ultimately spoil them by rendering them opake*."

To those who will read a book on this subject, it is quite unnecessary to enter upon a formal vindication of the study of the lower tribes of organized life. I shall content myself with the observation that, whilst this study is not wanting in those qualities which recommend the other branches of Natural History as a means of intellectual improvement, it has a peculiar claim on account of the light which it reflects on the ultimate organization of living bodies in general. Physiologists have of late pursued the investigation of the structure and development of cells, both animal and vegetable, with the greatest zeal; and with good reason, since it is obviously most desirable that we should thoroughly understand these elements of organization before we attempt to explain its more complicated arrangements. For the attainment of so important an object, the Desmidieæ furnish the most valuable assistance. If the view which I have taken be correct, their frond, in most of the genera at least, consists of a single cell, which, although it is certainly more complex than cells in general, enables us to trace its own history with ease and certainty, and reveals to us that of forms still more simple.

It may suffice thus briefly to show the utility of such inquiries; but the improvements which the microscope has received in modern times may well suggest a wider application of the language in which the illustrious Ray vindicated his favourite pursuits†.

---

* Whilst my best thanks are due to Mr. Topping for this account of his mode of preparing the cells, I must observe that the neat execution of them requires more skill and leisure than most persons possess, and therefore presume I shall render an acceptable service to my readers by mentioning that cells of any shape or size that may be required, and also the thin glass necessary for closing them, are supplied by Mr. Topping. In justice to Mr. T., whose cells I have used extensively, I must bear testimony to their beauty and utility. His address is No. 1, York Place, Pentonville, London.

† "Non deerit qui me vanæ curiositatis arguat, quòd res adeò viles et ab-

What the Allwise did not disdain to create cannot be unworthy of our notice; and if in the minute Desmidieæ, so long concealed from the unassisted eye, we have been at length enabled to recognize objects as carefully organized as the bulky elephant or the majestic oak, and as happily adapted to their position in nature, possessing too an economy whose laws are no less constant and regular, shall we not gladly examine this fresh evidence of an Almighty hand, as distinctly impressed on them as on the rest of his creation?

"To Him no high, no low, no great, no small;
He fills, He bounds, connects and equals all."

---

jectas, nullius in vita usûs, indagaverim, iisque describendis tantum temporis et operæ impenderim. Cui respondeo, Quòd Dei opera sunt in quibus contemplandis memet exerceo; quòd Divinæ Artis et Potentiæ effecta, quibus exquirendis subsecivas horas addico; quòd Ille me in hunc mundum introduxerit, tam inexplicabili rerum varietate instructum et ornatum; quòd oculis, quos mihi contulit, ea videnda, animo consideranda objecerit. In Dei ergo contumeliam redundat, quòd hæc, quæ eum creâsse negare non audes, supervacua et inutilia esse affirmes."—*Hist. Plant.* v. 3. præf. p. ii.

SYNOPSIS OF THE GENERA.

Fam. DESMIDIEÆ.

FRESHWATER, figured, mucous and microscopic Algæ, of a green colour. Transverse division mostly complete, but in some genera incomplete. Cells or joints of two symmetrical valves, the junction always marked by the division of the endochrome, often also by a constriction. Sporangia formed by the coupling of the cells and union of their contents.

* Plant an elongated jointed filament. Sporangia orbicular, smooth.

1. Hyalotheca. Filament cylindrical.
2. Didymoprium. Filament cylindrical or subcylindrical; joints with two opposite bidentate projections.
3. Desmidium. Filament triangular or quadrangular; joints connected by a thickened border.
4. Aptogonum. Filament triangular or plane, with foramina between the joints.
5. Sphærozosma. Filament plane, margins incised or sinuated; joints with junction glands.

** Frond simple from complete transverse division, distinctly constricted at the junction of the segments, which are seldom longer than broad. Sporangia spinous or tuberculated, rarely if ever smooth.

6. Micrasterias. Lobes of the segments incised or bidentate.
7. Euastrum. Segments sinuated, generally notched at the end, and with inflated protuberances.

8. Cosmarium. Segments in front view neither notched nor sinuated, in end view elliptic, circular or cruciform.

9. Xanthidium. Segments compressed, entire and spinous.

10. Arthrodesmus. Segments compressed, and having only two spines or mucros.

11. Staurastrum. End view angular, radiate, or with elongated processes, which are never geminate.

12. Didymocladon. Segments angular, each angle having two processes, one inferior, and parallel with the similar one of the other segment, the other superior and divergent.

*** Frond simple from complete transverse division, generally much-elongated, never spinous, frequently not constricted at the centre. Sporangia smooth.

13. Tetmemorus. Frond straight, constricted at the centre and notched at the ends.

14. Penium. Frond straight, scarcely constricted at the centre.

15. Docidium. Frond straight, much-elongated, constricted at the centre, truncate at the ends.

16. Closterium. Frond crescent-shaped or arcuate, not constricted at the centre.

17. Spirotænia. Frond straight, not constricted at the centre; endochrome spirally twisted.

**** Cells elongated, entire, fasciculated.

18. Ankistrodesmus. Cells aggregated into faggot-like bundles.

***** Frond composed of few cells, definite in number, and not forming a filament. (Sporangia unknown.)

19. Pediastrum. Cells arranged in the form of a flattened star; their outer margin bidentate.

20. Scenedesmus. Cells oblong or fusiform, entire, placed side by side in a single row, but during division in two rows.

BRITISH DESMIDIEÆ.

* *Plant an elongated jointed filament.   Sporangia orbicular, smooth.*

1. HYALOTHECA, *Ehr.*

Filament elongated, cylindrical, very gelatinous; *joints* having either a slight constriction which produces a crenate appearance, or a grooved rim at one end which forms a bifid projection on each side. *End view* circular.

The filaments are cylindrical, simple, jointed, invested with a broad gelatinous sheath, and very fragile in one species but not so in the other. Either a groove passes round each joint, giving a crenate appearance to the margins of the filament and dividing the endochrome into two portions, or else a grooved rim at one extremity of the joint appears on each side like a bifid process.

A transverse view shows a radiate endochrome in one species; but I have been unable to determine whether this occurs in the other, on account of its imperfect fragility.

The cylindrical filament distinguishes this genus from *Desmidium* and *Sphærozosma*. It has no angular projections, is not twisted, and has always the same apparent breadth, and consequently in all these respects differs from *Didymoprium*.

1. *H. dissiliens* (Smith); filament fragile, crenate; a shallow groove round each joint dividing the endochrome into two portions.

Conferva dissiliens, Smith, *Eng. Bot.* t. 2464. (1812) (not Dillw.).
Desmidium mucosum, Bréb. *Alg. Fal.* p. 65. t. 2 (1835). Menegh. *Synop. Desmid. in Linnæa* 1840, p. 204.   Ralfs *in Annals of Nat. Hist.* v. 11. p. 374. t. 8. f. 2.
Desmidium limbatum, Chauv. *Alg. de la Norm.* f. 6. (1836), according to Meneghini.

Glœoprium dissiliens, Berk. *in Annals of Nat. Hist.* v. 16. p. 11 (1845). Jenner, *Fl. of Tunbridge Wells*, p. 192. Hassall, *Brit. Freshwater Algæ*, p. 346. t. 83. f. 3.
Hyalotheca mucosa, Kütz. *Phycologia Germanica*, p. 140? (1845).
Hyalotheca dissiliens, Bréb. *in lit.* (1846).

Common. Henfield, Sussex; and Cwm Bychan, North Wales, *Mr. Borrer*. Swansea; Dolgelley; Bedgelert; Penzance, &c., *J. R.* Hampshire and many stations in Sussex, Kent and Surrey, *Mr. Jenner*. Essex and Herts, *Mr. Hassall*. Ayrshire, *Rev. D. Landsborough*. Aberdeen, *Dr. Dickie* and *Mr. P. Grant*. Bandon, *Professor Allman*. Kerry, *Mr. Andrews*. Ambleside, *Mr. Sidebotham*. Yate near Bristol, *Mr. Broome*. Stoke Hill near Wells, *Mr. Thwaites*. Rochdale, *Mr. Coates*. Manchester, *Mr. Williamson*.

Germany, *Kützing*. Falaise, *Brébisson*. West Point, New York, *Professor Bailey*.

Filaments mucous, adhering firmly to paper. The joints are usually broader than long, and as each has a shallow groove passing round it, it resembles a small pulley-wheel, and the filament acquires a crenate appearance; the crenatures are generally very shallow, sometimes nearly obsolete, but I have occasionally seen them deeper, when the plant looks more like the figure in ' Eng. Bot.;' at all times they may be detected on a careful examination with the higher powers of the microscope. The endochrome is divided into two portions by this constriction, and there is scarcely any depression between the joints.

The mucous sheath is easily observed, and on each side of the filament is nearly as broad as the central coloured portion. The sheath and filament are alike cylindrical.

If kept in water for a few days the plant separates into single joints, each of which has a perfect mucous covering. In the Desmidieæ in general the filaments separate into single joints with a facility proportioned to the greater development of the sheath; but *Hyalotheca mucosa* in this, as in other respects, deviates from the rest of the group. The breaking up takes place more speedily in the present plant, in *Didymoprium Grevillii* and *Sphærozosma vertebratum*, than in *Didymoprium Borreri* or in the genus *Desmidium*. This circumstance seems to depend upon the elasticity of the mucous covering of the joint. This elasticity is proved by its becoming longer than the joint immediately upon separation, and hence into whatever number of pieces a filament may be broken, the ends of each portion are as perfectly covered as the original end of the filament itself, and with regard to each particular joint the result is the same.

The transverse view is circular, and shows a mucous border of the same form. In this aspect the endochrome is generally disposed in a stellate manner, with six or seven rays, and frequently has a colourless central spot.

Foreign specimens of " *Desmidium mucosum*," given me by the Rev. M. J. Berkeley, agree in all points with the above description.

Hyalotheca dissiliens frequently occurs in considerable quantities, unmixed with other Algæ. When young it agrees in appearance with many of the Conjugatæ, but it may be distinguished by its fragility. It soon becomes of a pale opake green, and is then even more fragile than before.

Although I had seen sporangia of *H. dissiliens* at Penzance, I was unable to determine to what plant they belonged until Mr. Jenner sent them from Sussex and suggested that they were its conjugated state. At first the difficulty I experienced in recognizing the original cells in their altered condition induced me to hesitate in admitting the correctness of his opinion. But having subsequently gathered this species in all stages near Dolgelley, I have traced its changes, with Mr. Jenner's assistance, and satisfied my doubts.

The filaments separate into single joints, which couple by tubes; so far the manner of conjugation is similar to the examples I had previously seen in *Staurastrum* and *Tetmemorus*, but in those genera the segments of the cell are completely divided by the growth of the newly-formed portion; in the present case the cells separate only on the sides opposed to each other. In the cleft is formed the new portion, the enlargement of which, after it has become the connecting tube, by causing the cell to gape, produces a notch-like appearance on the outer margin. The process would be more evident if the similar texture of the old and new parts did not render it difficult at first sight to detect the joint in its altered form; and the difficulty is increased, as the coupling usually takes place in a crossed or twisted position, which thus still farther disguises the joints.

The sporangium, which is circular, is situated within the connecting tube, and the empty joints are permanently attached to it.

I have been unable to determine, by my own observation, whether the mucous covering remains on the conjugated cells, but Mr. Jenner informs me that in some instances he has observed it, and indeed that its presence led him to the identification of the plant. As in most instances, however, it is either wanting or very obscure, I have thought it advisable to omit it in the figures of the sporangia, and merely to notice its occasional occurrence.

Sporangia have been gathered also by Mr. Thwaites, Dr. Dickie, Mr. Broome, and Mr. Sidebotham; and M. de Brébisson informs me that he has met with them in France.

Length of joint from $\frac{1}{2105}$ to $\frac{1}{1351}$ of an inch; breadth of filament from $\frac{1}{1308}$ to $\frac{1}{833}$; breadth of sheath from $\frac{1}{647}$ to $\frac{1}{357}$. Diameter of sporangium $\frac{1}{825}$.

Tab. I. f. 1. *a, b.* portions of mature filaments; *c.* portion of a dividing filament; *d.* empty joint; *e, f.* transverse view; *g.* cells conjugating; *h, i.* sporangia.

2. **H. mucosa** (Mert.); filament scarcely fragile; joints not constricted, but having at one of the ends a minute bidentate projection on each margin, the adjoining end of the next joint being similar.

Conferva mucosa, Mert. Dillw. *Brit. Conf.* t. B. (1809). Hook. *Br. Fl.* v. 2. p. 351. Harv. *Br. Alg.* p. 127.

Hyalotheca mucosa, Ehr. *Kurze Nachricht über 274 seit dem Abschluss der Tafeln des grössern Infusorienwerkes?* (1840).

Glœoprium mucosum, Ralfs, *Annals of Nat. Hist.* v. 16. p. 11. t. 3. f. 6. (1845); *Trans. of Bot. Soc. of Edinburgh*, v. 2. p. 165. t. 18. Jenner, *Fl. of Tunbridge Wells*, p. 192. Hass. *Br. Freshwater Algæ*, p. 346. t. 83. f. 5.

Hyalotheca Mertensii, Brébisson *in lit.* (1846).

In shallow pools and gently-flowing streams; most plentiful in the autumn. Bantry, *Miss Hutchins*. Appin, *Capt. Carmichael*. Trentishoe, Devonshire; Penzance; and Dolgelley, *J. R.* Herts and Essex, *Mr. Hassall*. Chiltington Common, near Pulborough, Sussex; Ashdown Forest; and in the peat-bog at Fisher's Castle, near Tunbridge Wells, *Mr. Jenner*. Near Aberdeen, *Mr. P. Grant*.

Falaise, France; *Brébisson*.

Filaments elongated, very gelatinous, of a pale translucent green; not fragile. Under the microscope the joints are generally about equal in length and breadth, and the endochrome forms a single irregular patch. The joints are not constricted, but at one end they have on each margin a minute bidentate projection: as the similar ones of the next joint are at its adjoining extremity, these projections occur only near the alternate dissepiments. When however the joint is elongated, preparatory to the formation of two joints, the endochrome is divided into two portions, and then these projections are present at both ends, the next joint undergoing a similar change. These projections are extremely minute, and can be detected only by employing the highest powers of the microscope; and even then are liable to be overlooked, if not carefully sought for. I examined many specimens of this plant in 1841, but did not perceive these curious protuberances until the following year. I believe they are really formed by a grooved rim round the end of the joint, because however the filaments may be moved, they are equally apparent; whereas if they were real processes, as in *Didymoprium*, they would be sometimes either entirely concealed or rendered less distinct; they can also be traced like a transverse band across the empty cell.

The filaments have a very broad mucous sheath, which from its great breadth and absence of colour is not easily discerned; it is more evident when a specimen is dried on talc or glass, as the margins are then generally perceptible. When first gathered the filaments are very distinct, frequently parallel and subdistant even to the naked eye: this depends on the great breadth of their mucous sheaths, which prevents the coloured filaments from coming into contact. By this character *H. mucosa* may in general be known, even without the aid of a microscope. The same circumstance occurs in young plants of *H. dissiliens*, but is less remarkable, as its mucous sheath is not more than half as broad.

Under a low power of the microscope, *H. mucosa* has considerable resemblance to *H. dissiliens*, with which it is probably not unfrequently confounded. But they may always be distinguished, even without the aid of a microscope. *H. dissiliens* is extremely fragile, and will break into pieces if a small portion be placed on the hand and the finger gently passed over it; this plant, on the contrary, will not break if it be taken out of the water and allowed to hang down in long strings. It always has a clear translucent appearance; but *H. dissiliens*, except when very young, is of an opake green. *H. dissiliens*, if kept in water for a few days, spontaneously separates into single joints; *H. mucosa*, although treated in the same manner for weeks, did not separate into fragments sufficiently small to enable me to obtain a satisfactory transverse view. At length, however, I was more fortunate. When kept in water the mucous sheath is partially dissolved, and interferes with the examination of the joints, but by repeatedly changing the water I succeeded in procuring many single joints. The transverse view was circular, and when the cell was empty the border appeared decidedly striated. The endochrome had some tendency to a radiate form, but the treatment to which the filaments had been subjected prevented me from forming any certain conclusion as to its disposition in the recent plant.

Under the microscope *H. mucosa* may be known from *H. dissiliens* by its joints never appearing crenate, and by its endochrome being almost always in a single patch, or by the greater length of the joints if the endochrome is bipartite. Its mucous sheath is with difficulty detected, and when seen will be found to extend on each side twice the breadth of the coloured filament; whereas in *H. dissiliens* the mucous sheath is, except in old specimens, detected without difficulty.

The joints seem to be in pairs; a single one is consequently unsymmetrical.

H. mucosa agrees with the other Desmidieæ in its capability of being kept a long time without undergoing decomposition.

This plant differs in many respects from the other Desmidieæ; indeed so much so, that I had some doubts whether it would be correctly placed in this family; but as the Rev. M. J. Berkeley, as well as every other algological friend whose opinion I have solicited, considers that its place must be in the same genus with the preceding species, I have described it here.

I was first indebted to Mr. Hassall for the information that the plant under consideration is the *Conferva mucosa*, Dillw., as also for an opportunity of examining a foreign specimen under that name from the herbarium of Dr. Greville. I have since been enabled to compare our plant with a portion of an Irish specimen of *Conferva mucosa* presented to me by Mr. Borrer, who received it from Sir W. J. Hooker. From the latter I learn that this was an original specimen from Miss Hutchins. Sir W. J. Hooker has also presented me with an Appin specimen, collected by Capt. Carmichael. All these are identical with the present plant.

Length of joint from $\frac{1}{1666}$ to $\frac{1}{1250}$ of an inch; breadth of filament $\frac{1}{1250}$ to $\frac{1}{1111}$; breadth of sheath $\frac{1}{167}$.

Tab. I. f. 2. *a.* portion of a filament; *b.* empty joints; *c, d.* transverse views.

2. DIDYMOPRIUM, *Kütz.*

Filament elongated, gelatinous, fragile, regularly twisted, cylindrical, with a bidentate process or angle on each side of the *joint*.

The filaments are elongated, simple, jointed, gelatinous and very fragile, and finally separate into single joints; each joint has two opposite bidentate angles or processes. Hence the margins of the filament are crenate; and as it is regularly twisted, it not only appears of unequal breadth, but the form of its joints also varies, as more or less of the angles is seen at the margin; in short, as they are at one time fully visible, and at length entirely disappear.

In a transverse view the joints are circular or broadly elliptic, with two minute opposite projections which are formed by the angles. The endochrome is radiate; its rays from four to seven.

Conjugated specimens of both species have been observed, and possess much interest from their being the representatives in this family of analogous states in some of the Conjugatæ.

Didymoprium differs from *Desmidium* in having only two angles; in a transverse view also the number of rays of the endochrome does not depend on the number of angles.

Kützing separated this genus from *Desmidium* and *Hyalotheca*, and as this separation meets with the approbation of the Rev. M. J. Berkeley, I am induced, in deference to their joint opinion rather than from my own conviction of its necessity, to adopt it in the present work. Mr. Hassall is averse to its adoption; but surely he is mistaken when he puts the differential characters of *Didymoprium* and *Desmidium* on the same footing as the presence or absence of a mucous sheath, and thinks that if *Didymoprium* is a good and valid genus, the absence of a mucous sheath would be sufficient to separate from it *Didymoprium Borreri*.

The absence or presence of a mucous sheath, or rather its comparative development, since traces of it may be found in all the filamentous Desmidieæ, can only be of specific value; and Mr. Hassall himself, in the analogous example of the Conjugatæ, amongst which species of *Tyndaridea* and *Mougeotia* have equally developed sheaths, has only in one instance thought it necessary even to allude to its occurrence.

In addition to the characters already enumerated, I may observe that in *Desmidium* a transverse view shows that the cell itself, irrespective of the bidentate projections, is angular, and that the endochrome is divided into a number of rays corresponding with the number of angles: neither of these circumstances occurs in *Didymoprium*.

The filaments in this genus increase in length by the repeated division of the joints exactly as in the other genera, the new portions being formed between the original segments, which in other respects remain unaltered; the teeth of the angles furnish a guide to the change that takes place, for they become more widely separated, and then the new teeth appear, very minute at first, but gradually enlarging, until the new segment equals the older one in breadth.

1. *D. Grevillii* (Kütz.); sheath distinct; joints of the filament broader than long, with a thickened border at their junction; transverse view broadly elliptic.

Desmidium cylindricum, Greville, *Scot. Crypt. Fl.* t. 293. (1827); *in Hook. Brit. Fl.* v. 2. p. 402. Ag. *Consp. Diatom.* p. 56. Kützing, *Synop. Diatom. in Linnæa* 1833, p. 614. Brébisson, *Alg. Fal.* p. 64. Harv. *Brit. Alg.* p. 197. Menegh. *Synop. Desmid. in Linnæa* 1844, p. 204. Ralfs, *Annals of Nat. Hist.* v. 11. p. 373. t. 8. f. 1; *Trans. of Bot. Society of Edinburgh*, v. 2. p. 6. t. 2. Hassall, *Brit. Freshwater Algæ*, p. 342. t. 83. f. 1.

Arthrodesmus? cylindricus, Ehr. *Infus.* p. 142 (1838).

Hyalotheca cylindrica, Ehr. *Kurze Nachricht über 274 seit dem Abschluss der Tafeln des grössern Infus. neu beobachtete Infus.* (1840).

Desmidium compressum, Corda, *Obser. micros. sur les Anim. de Carlsbad*, p. 18 (1840).

Didymoprium Grevillii, Kützing, *Phycologia Generalis*, p. 166 (1843); *Phy. Germ.* p. 141.

Didymoprium cylindricum, Ralfs, *Annals of Nat. Hist.* v. 16. p. 10 (1845); *Trans. of Bot. Soc. of Edinburgh*, v. 2. p. 164.

Hyalotheca Grevillii, Brébisson *in lit.* (1846).

Appin, *Capt. Carmichael*. Dolgelley; Carnarvon; Penzance, &c., *J. R.* Chiltington Common near Pulborough, and Ashdown Forest, Sussex; near Southampton, Hants, and Reigate, Surrey, *Mr. Jenner*. Cheshunt, *Mr. Hassall*. Meath, *Mr. Moore*. Aberdeenshire, *Dr. Dickie* and *Mr. P. Grant*. Manchester, *Mr. Williamson*. Ambleside, *Mr. Sidebotham*.

Prussia, *Kützing*. Prague; Carlsbad; Reichenberg, &c., *Corda*. Falaise, *Brébisson*. New York; Rhodes Island, *Bailey*.

The filaments in *Didymoprium Grevillii* are as thick as those of *Desmidium Swartzii*, which in water is a very similar plant, and are very gelatinous and fragile; their colour is pale green, but, except when very young, somewhat opake. The joints are connected by a thickened border, and, inclusive of the angles, which are colourless, are rather broader than long, and oval; the angles are bidentate, their teeth angular. When the angles are not visible the joints are nearly quadrate.

The mucous sheath is very evident, narrower in this plant than in *Hyalotheca dissiliens*, and waved at the margin. When the angles are not visible the undulation of the margin is slight, but becomes more marked as the angles become visible. The sheath is jointed as well as the filament, the lines of separation being sometimes distinctly apparent.

The filaments are regularly twisted, and at about every tenth joint the angles become very perceptible, whilst in the two central joints they are almost invisible; on slightly altering the position of the filament the angles in the latter become visible, and in the former disappear.

On account of the oval form of the joints their respective angles are separated, and thus the filament acquires a pinnatifid appearance.

The transverse view is broadly elliptic, with a small process at each end, representing the angles. The mucous sheath is of the same form, and protuberant over the processes. The endochrome is four- or five-rayed.

I have gathered the conjugated state of *Didymoprium Grevillii* near Dolgelley. It has been found in France by M. de Brébisson, but I believe that the present is the first description of it.

Before coupling *D. Grevillii* separates into single joints (in only one instance have I observed two joints of one filament still united and conjugated with two other separated joints); the cells gape at the margins opposed to each other, exactly as described under *Hyalotheca dissiliens*, but to a smaller extent, and the original joint is far more easily detected, as one of the teeth of the angle bounds the cleft on each side. As the cleft is narrower, there is an absence of the notch-like appearance so evident on the outer margin of *Hyalotheca dissiliens*. The cells become connected by a narrow process, often remarkable for its length, and the contents of one cell passes through it into the other, and a sporangium is formed in the same manner as in many of the Conjugatæ. The transfer of the endochrome takes place in a mass, and one part may frequently be observed uniting with that in the receiving-cell whilst the other is still in the tube. The joints couple in a crossed position, as in *Hyalotheca dissiliens*, but the length of the connecting tube permits them to remain apart. The joint which contains the sporangium is but little altered in a transverse view, and when the empty cell becomes detached can scarcely be distinguished, except by contrasting the denseness of the endochrome with its stellate appearance in the unconnected joints. This detachment of the empty cell is a frequent occurrence after the process is completed, nd forms another character of resemblance to those Conjugatæ which bear the sporangium within the cell.

In a front view the sporangium is as orbicular as the quadrate form of the joint will permit, being of course rather broader at the side which is cleft, and on that side is usually slightly protuberant. The mucous covering remains unaltered on the joints when they are coupled.

Length of joint $\frac{1}{464}$ of an inch; breadth of filament including teeth $\frac{1}{473}$; breadth of filament when the teeth are hidden, from $\frac{1}{633}$ to $\frac{1}{612}$; breadth of sheath from $\frac{1}{409}$ to $\frac{1}{312}$. Diameter of sporangium $\frac{1}{573}$.

Tab. II. *a.* portion of a filament; *b.* dividing joint; *c.* separated joints; *d.* transverse view; *e, f, g.* conjugated cells; *h, i, k.* sporangia.

2. **D. Borreri** (Ralfs); joints inflated, barrel-shaped, longer than broad; transverse view circular.

Desmidium Borreri, Ralfs, *Annals of Nat. Hist.* v. 11. p. 375. t. 8. f. 4. (1843); *Trans. of Bot. Society of Edinburgh*, v. 2. p. 8. t. 2. Hassall, *Brit. Freshwater Algæ*, p. 343. t. 83. f. 9.

Didymoprium Borreri, Ralfs, *Annals of Nat. Hist.* v. 16. p. 10 (1845); *Trans. of Bot. Society of Edinburgh*, v. 2. p. 164. Jenner, *Flora of Tunbridge Wells*, p. 192.

Bambusina Brebissonii, Kützing, *Phycologia Germanica*, p. 140 (1845).
Hyalotheca bambusina, Brébisson *in lit.* (1846).

Cwm Bychan, *Mr. Borrer.* Dolgelley ; Llanberris ; Tal Sarn near Lampeter, *J. R.* New Forest, Hants ; near Battle ; Ashdown Forest and Chiltington Common, Sussex, *Mr. Jenner.* Ireland, *Mr. Andrews.* Ambleside, Westmoreland, *Mr. Sidebotham.* Manchester, *Mr. Williamson.* Glen Lin, *Dr. Dickie.* Near Aberdeen ; and Moss Hagg, Banffshire, *Mr. P. Grant.*

Falaise, *Brébisson.* Germany, *Kützing.* Rhodes Island, *Bailey.*

Filaments pale green, soon becoming slightly opake, very slender, their mucous sheath wanting or indistinct. In the water this species resembles *Hyalotheca dissiliens,* but it is less slippery to the touch, and by this character may often be known when unmixed with other Algæ. The filaments are regularly twisted and complete a turn in about 16 joints. The latter are generally nearly twice as long as broad, but in some specimens their length scarcely exceeds their breadth ; they are inflated, so that when the angles are not seen, they resemble small barrels placed end to end, but when the angles are fully displayed their appearance may be compared to the juxtaposition of flower-pots by their mouths, the rims and intervals between which will represent the crenate angles. The angles are bicrenate, and the crenatures being minute and rounded, seem less like an interruption of the outline than in *D. Grevillii,* but rather as if the angles were attached to the sides of the joints. On account of the length of the joints, the disposition of the endochrome in two portions is very distinct. The joints have not a thickened border at their junction as in *D. Grevillii,* and the filament separates with less facility into single joints. Occasionally, especially in the short-jointed variety, faint longitudinal lines interrupted at the suture may, on careful examination, be perceived, but I find them so uncertain and indistinct, even under the triplet, that I consider it advisable merely to allude to their occurrence ; I have however introduced a figure showing them as Mr. Jenner has observed them in favourable specimens with his powerful achromatic instrument. Mr. Jenner, to whom the credit of their discovery is due, does not find them in every specimen, and regards them as affording an excellent test of the power of the microscope. He informs me that they are situated upon the inner surface of the cell. The transverse view is circular, with two minute opposite projections. The endochrome, stellate as in the other species, has five or six rays.

The conjugated state of this species I have gathered sparingly at Dolgelley. We have seen that in this family *D. Grevillii* is the only known example which has the spore formed within the cell. *D. Borreri* presents us with another point of analogy between the Desmidieæ and the Conjugatæ. Most of the former have the frond composed of a single cell, or the filament before conjugating becomes altogether broken up ; this is but imperfectly the case here, as the coupling cells have generally other cells remaining, attached either in an unaltered state, or conjugated on their part with portions of other filaments, and thus a kind of network is sometimes formed ; the coupling of the cells crossways occasions a still more confused appearance and renders it exceedingly difficult to trace them. The process is similar to that described under *Hyalotheca dissiliens*, but the segments are still more widely separated on the inner side, and consequently so far thrown back as to cause the attached joints to form a zigzag. The elliptic sporangium lies between the cells, which remain permanently attached to it.

Kützing has established for this plant a distinct genus (*Bambusina*). The characters which he relies upon are the absence of a mucous sheath, the cask-like joints, and their union, not by a thickened rim as in *Didymoprium Grevillii*, but in the manner of the Zygnemata. To these characters, if he had been acquainted with the reproductive state of the two plants, he would doubtless have added the different position of their sporangia.

Even if the presence of a sheath could be supposed to have a generic value, it would not apply to the present question; for, as Mr. Jenner has pointed out to me, a mucous covering does exist, and may occasionally be detected as a narrow border to the filament. The agreement also between the plants in their angles and the regular twisting of their filaments appear to me to forbid their separation.

Kützing suggests* that this plant may be identical with the *Gymnozyga moniliformis*, Ehr. (Bericht d. Berl. Ac. Novb. 1840). I am unable to determine this point, as I have seen no other notice of Ehrenberg's plant.

This plant was known to M. de Brébisson before it was described in the 'Annals of Natural History,' and indeed under the name of *Desmidium bambusinum* was included by him in a list of Desmidieæ inserted in a treatise on the microscope by M. Chevalier in 1839, but I have retained the present name, it being the first under which it was either described or figured.

Length of joint $\frac{1}{939}$ of an inch; breadth of the joint including teeth $\frac{1}{1031}$. Length of sporangium $\frac{1}{880}$, breadth $\frac{1}{1106}$.

Tab. III. *a, b.* portions of mature filaments; *c.* portion of a filament with joints dividing; *d, e, f.* portions of filaments without endochrome to show the longitudinal markings; *g, h.* transverse views; *i, k.* portion of filaments conjugating; *l, m.* sporangia with the empty fronds attached.

3. DESMIDIUM, *Ag.*

Filament fragile, elongated, triangular or quadrangular, regularly twisted; *joints* bidentate at the angles.

The filament is fragile, of a pale green colour and slightly opake. When dried, the British species usually acquire a yellowish appearance, and adhere to paper or talc less firmly than plants belonging to allied genera. They are regularly twisted, but being triangular or quadrangular, two of the bidentate angles of each joint are always visible at the margins. The endochrome is divided into linear portions by a pale transverse line between the angles.

---

* Phycologia Germanica, p. 140.

Traces of a mucous sheath have been detected by the Rev. M. J. Berkeley and more recently by myself.

A transverse view shows that the cell is triangular or quadrangular, and that the endochrome has thick rays corresponding in number with the angles; these rays are frequently cloven.

Recent specimens of *Desmidium* are easily known from other genera by one or two dark waved lines passing down the filament, which appearance is occasioned by the twisting of the angular filament.

1. *D. Swartzii* (Ag.); filament triangular, equal, with a single longitudinal, waved, dark line formed by the third angle; end view triangular with the endochrome three-rayed.

Diatoma Swartzii, Ag. *Disp. Alg. Svec.* (1811). Lyngb. p. 177. t. 61.
Desmidium Swartzii, Ag. *Syst.* p. 9 (1824). *Consp. Crit. Diat.* p. 56. Greville, *Scot. Cryp. Fl.* t. 292. Kützing, *Synopsis Diatom. in Linnæa* 1833, p. 613; *Phycologia Germ.* p. 141. Greville *in Hook. Brit. Fl.* v. 2. p. 402. Brébisson, *Alg. Fal.* p. 53. t. 2. Ehrenberg, *Infus.* p. 140. t. 10. f. 8. Meneghini, *Synop. Desmid. in Linnæa* 1840, p. 203. Bailey, *American Bacillaria in Amer. Journal of Science and Arts*, v. 41. p. 288. t. 1. f. 1. Harvey, *Manual of Brit. Algæ*, p. 196. Corda, *Observ. Microscop. sur les Animal. de Carlsbad*, p. 17. Ralfs, *Annals of Nat. Hist.* v. 11. p. 375. t. 8. f. 3; *Trans. of Bot. Soc. of Edinburgh*, v. 2. p. 7. t. 2. Hassall, *Brit. Freshwater Algæ*, p. 344. t. 83. f. 7.

Common. Appin, *Capt. Carmichael.* Tunbridge Wells, *Mr. Borrer.* Swansea; Carnarvon; Dolgelley; Penzance, &c., *J. R.* Many stations in Sussex, and near Southampton, *Mr. Jenner.* Essex, *Mr. Hassall.* Caragh Lake, Kerry, *Mr. Andrews.* Meath, *Mr. Moore.* Westmoreland, *Mr. Sidebotham.* Rochdale, *Mr. Coates.* Manchester, *Mr. Williamson.* Aberdeenshire, *Dr. Dickie* and *Mr. P. Grant.*

Sweden, *Agardh.* North Germany, *Martens, Kützing* and *Ehrenberg.* Carlsbad; Prague; Reichenberg, *Corda.* Throughout the United States, *Bailey.* Falaise, *Brébisson.*

The filaments are elongated and very fragile, but separate spontaneously into single joints less readily than in *Hyalotheca dissiliens* and *Didymoprium Grevillii*. They adhere but slightly to paper. They are triangular, and, when viewed under the lens, the margins are always formed by two of the three angles in turn, whilst the dark line regularly passing from side to side marks the third angle and shows that the filament has a spiral twisting in about 16 joints. On examining the filament where the dark line touches the margin, if we first raise and then depress the lens, both angles at the point of apparent contact may be distinctly seen.

The joints are in the front view somewhat quadrangular, broader than long, and each angle has two minute, slightly angular teeth. The notch between the teeth is very distinct but not gaping. The joints are connected by a thickened margin, which partly fills the notch formed between the joints by the projection of the angles, whence the filament has a pinnatifid appearance.

The transverse view is triangular; the angles are blunt and the sides slightly concave. The endochrome is in three portions, placed in the angles and connected together at the centre; but this arrangement of the endochrome is seen only when the joints separate spontaneously. The rays are usually bipartite. When a joint is separated by force under the microscope, a cloud of minute granules is poured out and for an instant obscures the view.

Mr. Borrer has kindly presented me with a portion of a specimen of the plant here described, which was given him by Mr. Dawson Turner as the *Conferva dissiliens* of Dillwyn.

I have gathered at Dolgelley some fragments of this plant which had the endochrome condensed into a sporangium-like body in the centre of each joint. As in every other species of this family in which the reproductive body has been detected, it is the result of the coupling of the cells, I think it best merely to direct attention to the fact I have mentioned, leaving its nature to be determined by future observation.

Length of joint from $\frac{1}{2000}$ to $\frac{1}{1666}$ of an inch; breadth of filament $\frac{1}{633}$.

Tab. IV. *a*. portion of a mature filament; *b*. portion of a filament with the joints dividing; *c, d, e.* transverse views; *f.* sporangia?

2. **D. quadrangulatum** (Ralfs); filaments quadrangular, varying in breadth from the twisting of the filament and having two longitudinal waved lines; the end view quadrangular with the endochrome four-rayed.

Desmidium quadrangulatum, Ralfs, *in Annals of Nat. Hist.* v. 15. p. 405. t. 12. f. 9. (1845); *Trans. of Bot. Soc. of Edinburgh*, v. 2. p. 163.
Desmidium quadrangulare, Kütz. *Phy. Germ.* p. 141 (1845).

Hab. In a boggy pool at Bologas near Penzance; Dolgelley, *J. R.* Ambleside, *Mr. Sidebotham.* Aberdeen, *Dr. Dickie* and *Mr. P. Grant.*

The Harz, Germany, *Kützing.*

Mr. Berkeley and Mr. Borrer regard this plant as a variety of *D. Swartzii*, and certainly all the more obvious distinctive marks depend upon the filament being quadrangular in one case and triangular in the other*. As I have gathered it for two succeeding years quite unmixed with *Desmidium Swartzii*, and as Mr. Berkeley well observes (in a letter), that "whether considered as a species or variety, it is a remarkable plant and well-deserving of notice," I have preferred to describe it as distinct, although I do not consider the point free from doubt. As there are three sides in one plant and four in the other, whilst the sides in both are equal, the filament of *D. quadrangulatum* is stouter; for the same reason, instead of one dark longitudinal line it has two

---

* " Est-il réellement plus distinct du *D. Swartzii* que les Staurastr. à quatre rayons des Staur. à trois rayons?"—*Brébisson in lit.*

lines running from side to side and crossing each other : the additional line of course depends on the additional angle. When these lines approach the opposite margins of the filament only one side is presented to the eye, and the filament is then ,of the same breadth as in *D. Swartzii*, but as it is regularly twisted its apparent breadth varies, being greatest where both dark lines cross each other in the middle. The end view also has one angle more, and therefore the endochrome exhibits four instead of three rays. I may however observe, that the teeth, as Mr. Jenner has pointed out, are rounded in my specimens of *D. quadrangulatum* and angular in *D. Swartzii*; but further experience must determine whether this character is constant.

D. quadrangulatum requires about 40 joints to complete a turn, or before the same angle again appears at the same margin.

It affords some sanction to this arrangement, that shortly after my notice of this plant appeared in the 'Annals of Natural History,' it was described by Kützing in his 'Phycologia Germanica' as a distinct species.

Length of joint $\frac{1}{1244}$ of an inch; least breadth or side in transverse view $\frac{1}{603}$; greatest breadth, or diagonal, in transverse view $\frac{1}{455}$.

Tab. V. *a.* portion of a filament; *b.* filament less magnified; *c.* empty joint; *d, e.* transverse views.

4. APTOGONUM.

Filament elongated, triangular or plane; *joints* bicrenate at the margins; an oval foramen between the joints.

The filaments are elongated, jointed, fragile, and either triangular or plane; the joints at the angles are bicrenate, and at their junction are excavated at the centre, and thus form a series of oval foramina. The endochrome is bipartite.

By Brébisson, Meneghini and Kützing this plant is included in *Desmidium*, and certainly the state with triangular filaments does at first sight appear closely allied to *Desmidium Swartzii*; but the large oval foramen between the joints is so remarkable a character, that I must concur with Ehrenberg in placing it in a separate genus. As however Ehrenberg's name, *Odontella*, had previously been given by Agardh to some Diatomaceæ, it became necessary to find another name.

Ehrenberg included *Sphærozosma* in his *Odontella*, but the present genus is essentially distinct from *Sphærozosma*; in the latter the joints have their margins incised or sinuated, and gland-like processes at their junctions, and it is merely by the interposition of these processes that the joints are now and then slightly separated. In this genus, on the contrary, the joints have two prominent teeth at the margins or angles, and the foramen results from the excavation or concavity of the joint itself, and not from the presence of glands.

1. *A. Desmidium* (Ehr.); joints in the front view quadrangular, broader than long; crenatures distinct.

    α. Filaments triangular, regularly twisted; crenatures rounded.

Desmidium aptogonum, Bréb. *Alg. Fal.* p. 65. t. 2. (1835). Menegh. *Syn. Desm. in Linn.* 1840, p. 203. Kütz. *Phy. Germ.* p. 141.

    β. Filaments plane; crenatures shallower and slightly angular.

Odontella Desmidium, Ehr. *Infus.* p. 153. t. 16. f. 4. (1838).

    α. Ambleside, Westmoreland, *Mr. Sidebotham.* Dolgelley, *J. R.*

    β. Ambleside, *Mr. Sidebotham.*

North Germany, *Ehrenberg.* Carlsbad, *Corda.* Falaise, *Brébisson.*

The filaments, which are slenderer than those of *Desmidium Swartzii*, are slightly mucous, but have no distinct sheath. The joints are nearly as long as broad, and being excavated at their junction and connected only at the margins, they leave oval foramina between them.

The two states of this plant are so different that they seem at first sight distinct species. In α. the filaments are triangular, and being regularly twisted as in *Desmidium*, they in like manner exhibit a dark waved longitudinal line, marking the third angle; hence the foramina are far less conspicuous in this than in the plane variety, as they are obscured by the junctions of the third angle. When the dark line which denotes that angle approaches the margin of the filament they are more evident, and their size and figure may be ascertained. The marginal crenatures are deeper and more rounded, which fact, as Mr. Jenner has justly remarked to me, makes it doubtful whether a similar character can be relied on for distinguishing *Desmidium quadrangulatum* from *Desmidium Swartzii*. The end view is triangular; the sides are slightly concave and the angles rounded and somewhat thickened. The endochrome in this aspect has the rays equal in number to the angles, each ray being usually bipartite.

The variety β. is an elegant microscopic object. Its foramina are large and conspicuous. The only specimens I have seen were gathered by Mr. Sidebotham. Its characters are so distinct that it is unnecessary to compare this variety with any other plant.

A careless observer might mistake the triangular form for *Desmidium Swartzii*. It differs however not only in the presence of foramina, but its filaments are more slender and its joints longer in comparison with their breadth.

Several filamentous Desmidieæ have no slight resemblance to *Tæniæ*, and none more than this species.

Length of joint of α. (Dolgelley specimen), including connecting processes, $\frac{1}{1495}$ of an inch; length between foramina $\frac{1}{1818}$; breadth of joint $\frac{1}{1000}$.

β. (Ambleside specimen.) Length of joint, including the connecting processes, $\frac{1}{1295}$; length between foramina $\frac{1}{1938}$; breadth of joint $\frac{1}{925}$.

Tab. XXXII. f. 1. *a*. portion of mature filament; *b, c, d*. transverse views; *e*. portion of a filament of β. (Ambleside specimen); *f*. portion of a filament with dividing joints; *g*. two joints in an oblique position; *h*. lateral view.

5. SPHÆROZOSMA, *Corda*.

Filament plane, fragile; *joints* closely united by means of glandular processes, and deeply divided on each side, thus forming two segments, and giving a pinnatifid appearance to the filament.

The filaments are pale green, gelatinous, simple, plane, have a pinnatifid appearance from the division of the joints into two segments, are fragile, and finally separate into single joints. I have not observed that the filaments are twisted, as in *Desmidium* and *Didymoprium*. At the junction of the joints there are on each margin one or two minute glands or processes which are scarcely discernible in the front view, and do not interfere with the close junction of the joints. The transverse view is linear or oblong, and the processes, one or two at each side, are much more evident than in the front view.

This genus differs from *Desmidium*, *Didymoprium* and *Hyalotheca* in its flat filaments (which are not twisted), in the deep division of the joints into segments, and especially in the presence of the minute gland-like processes at the junction of the joints.

On account of its deeply constricted joints, this genus forms a connecting link between the preceding and following genera.

In *Sphærozosma*, as in the other genera with deeply constricted cells, the segments are frequently unequal during the growth of the plant, and they become in like manner equal when it approaches maturity, and its joints no longer divide.

1. *S. vertebratum* (Bréb.); joints as broad as long, deeply divided into two segments by a narrow notch on each side. Junction-glands oblique, solitary at the centre of each margin.

Desmidium vertebratum, Bréb. *Alg. Fal.* p. 65. t. 2. (1835).
Sphærozosma elegans, Corda, *Almanach de Carlsbad*, 1835, t. 4. fig. 37?; *Observ. Micros. sur les Animal. de Carlsbad*, p. 21. t. 4. f. 30? Hass. *Br. Alg.* p. 348. t. 84. f. 1.
Odontella? unidentata, Ehr. *Infus.* p. 159 (1838).
Isthmia vertebrata, Menegh. *Synop. Desmid. in Linnæa* 1840, p. 205.
Desmidium compressum, Ralfs, *in Annals of Nat. Hist.* v. 9. p. 253 (1842).

Schistochilum unidentatum, Ralfs, *in Jenner's Fl. of Tunbridge Wells*, p. 192 (1845).
Sphærozosma unidentatum, Ralfs, *in Annals of Nat. Hist.* v. 16. p. 14. t. 3. f. 7. (1845); *Trans. of Bot. Society of Edinburgh*, v. 2. p. 167.
Isthmosira vertebrata, Kütz. *Phycologia Germanica*, p. 141 (1845).

Hab. Chy-an-hâl and Kerris Moor near Penzance; Dolgelley, *J. R.* Rotherfield and near Tunbridge Wells, *Mr. Jenner.* Isle of Sheppey, *Mr. Bowerbank.* Ambleside, *Mr. Sidebotham.*

France, *De Brébisson.* Carlsbad, *Corda.* North Germany, *Ehrenberg* and *Kützing.* West Point, New York, *Bailey.*

The filaments are much compressed, and the joints, which are nearly equal in length and breadth, are so deeply constricted that at first sight a single one might be mistaken for two. This is more especially the case whilst the endochrome obscures the view of the union of the segments; as soon however as the joint becomes empty, its nature is distinctly seen. Though in other respects symmetrical, one segment is frequently much smaller than the other.

At the centre, where the joints are connected, is a minute gland or process arising from each margin. The projection of these glands is easily seen, even before the joints separate.

By a transverse view the joints are shown to be compressed, and the oblique glands, which are globular and supported on a very short stipes, are very distinct. In this view the joints are about twice as long as broad, with slightly concave sides, and the endochrome is stellate with from four to six rays.

In a recent state *Sphærozosma vertebratum* is very gelatinous and furnished with a broad mucous sheath, which from its tenuity and want of colour is very difficult of detection, and consequently I at first described it as absent. When the plant occurs unmixed with others, the presence of the sheath is indicated by the distinct and parallel filaments exactly as in *Hyalotheca mucosa*; for in both plants this appearance is produced by its great breadth, which prevents the closer contact of the coloured centres. The sheath is sufficiently apparent in specimens preserved on glass, and is on each side nearly as broad as the filament. In dried specimens it has an irregular waved margin and faint transverse markings.

The greater difficulty of detecting the sheath in recent specimens of *Hyalotheca mucosa* and *S. vertebratum* than in *Hyalotheca dissiliens* and *Didymoprium Grevillii*, seems to depend more upon its condition in the two former than upon its greater breadth. As in them it is less condensed, it is not only liable to be confounded with the water in which it is viewed, but is also more soluble in it. When *Didymoprium Grevillii* is mounted in fluid, the sheath remains as distinct as in recent specimens; the case is nearly the same when *Hyalotheca dissiliens* is so treated, but not so in respect of the present plant, the sheath of which seems to be dissolved.

Length of joint $\frac{1}{1429}$ of an inch; breadth of segments from $\frac{1}{909}$ to $\frac{1}{666}$; breadth at constriction $\frac{1}{2237}$; diameter of sporangium $\frac{1}{1250}$.

Tab. VI. f. 1. *a, b.* portions of filaments; *c.* empty joints; *d.* side view; *e.* glands in front view; *f, g.* transverse views. Tab. XXXII. f. 2. *a, b, c.* sporangia.

2. *S. excavatum* (Ralfs) ; joints longer than broad, having a deep sinus on both sides, and two sessile glands on each margin at their junction.

Schistochilum excavatum, Ralfs, *in Jenner's Fl. of Tunbridge Wells*, p. 192 (1845).

Sphærozosma excavata, Ralfs, *in Annals of Nat. Hist.* v. 16. p. 15. t. 3. f. 8. (1845) ; *Trans. of Bot. Society of Edinburgh*, v. 2. p. 168. Hass. *Brit. Freshwater Alg.* p. 349.

Hab. Pools. Dolgelley and Penzance, *J. R.* Cross-in-hand, and Ashdown Forest, Sussex; bogs at Fisher's Castle, Kent; Farnham, Surrey, and near Southampton, *Mr. Jenner.* Bristol, *Mr. Broome.* Ambleside, *Mr. Sidebotham.* Rochdale, *Mr. Coates.*

Falaise, *Brébisson.* West Point, New York, *Bailey.*

Very minute, seldom more than twenty-five joints in the filament, which is fragile and finally separates into single joints ; at their junction, in the front view, are two minute glands or processes, situated one near each angle, and nearly invisible before the escape of the endochrome. The joints are nearly twice as long as broad and much constricted in the middle ; the constriction is like an excavation or broad sinus on each side, so that the margins of the filament appear sinuated. The endochrome is pale bluish green with minute scattered granules.

The transverse view is oblong with four sessile glands, two on each side, and situated near the ends.

From their pale colour and minute size I have experienced much difficulty in determining the form of the angles, which in some specimens seem to me entire, but in others emarginate. Mr. Jenner, using a more powerful microscope, informs me that each is apparently furnished with three minute teeth, which can be detected only when favourably situated for observation. On each segment of the empty joint he finds three transverse series of minute granules, whose appearance at the margins produces this minutely toothed appearance.

I have frequently gathered this species, but always in small quantity, mixed with other Desmidieæ ; I am therefore unable to decide whether it possesses a mucous sheath.

The sporangia of *S. excavatum* have been gathered near Farnham in Surrey, and at Cross-in-hand, Sussex, by Mr. Jenner, who sent his specimen to me for examination. The joints after separation couple ; the sporangium is situated between them, and is large compared with the conjugating cells. It is elliptical, and the empty cells are closely connected with it, one at each end. As the specimen was mounted, I was prevented from using the triplet ; but from Mr. Jenner's description, the conjugation takes place from their flat or front surfaces, and not in the usual manner from one of the lateral sinuses ; therefore either one of the ends or one of the lateral margins will be presented to the eye.

Length of joints $\frac{1}{2575}$ of an inch ; breadth of segments $\frac{1}{3050}$; breadth at constriction $\frac{1}{5000}$; length of sporangium $\frac{1}{1562}$; breadth from $\frac{1}{2325}$ to $\frac{1}{1824}$.

Tab. VI. fig. 2. *a.* portion of a filament; *b.* portion of a filament with empty joints; *c.* empty joint to show the granules; *d.* transverse view; *e, f, g.* sporangia.

** *Frond simple (binate only when dividing), consisting of two segments united by a suture, where it is most frequently constricted.*

6. MICRASTERIAS, *Ag.*

Frond simple, lenticular, deeply divided into two, lobed segments; the lobes inciso-dentate (rarely only bidentate), and generally radiant.

The fronds are simple, nearly flat, without inflated protuberances, and divided nearly to the centre, so that the segments are semi-orbicular, and usually in close contact with each other along their entire breadth; each is deeply divided into lobes, which are generally arranged in a radiant manner and regularly cleft and dentate at the margin.

In two species sporangia have been detected; they are large, globular, and furnished with stout spines, which at first are simple, but finally become branched at the apex. In the perfect state they are particularly interesting from their resemblance to the fossil "Xanthidia" of Ehrenberg and others.

The orbicular, plane and deeply incised fronds will distinguish *Micrasterias* from all other genera in this family. In *Euastrum*, the only one with which it can be confounded, the fronds are oblong and the lobes are not incised.

As the plants to which Agardh affixed the name of *Micrasterias* are probably all included under the above description, I have followed Meneghini in retaining that name for this genus, especially as it seems to me that not one of Agardh's species is contained in the *Micrasterias* of Ehrenberg, who in fact intended by it a very different genus, the *Pediastrum* of Meyen*.

Although *Micrasterias* contains but few species, they admit of an arrangement in three sections.

The species in the first section present some of the most beautiful microscopic objects amongst the Desmidieæ. They are distinctly

---

* "Nomen a cl. Agardh propositum et perperam ab aliis Pediastris affixum, et cum novo illo Euastrum commutatum servandum."—*Menegh. Syn. Desmid. in Linnæa* 1840, p. 215.

visible, appearing to the naked eye like minute green dots. The microscope shows them to be circular and their segments deeply divided into five lobes. The end lobe, which is narrowest, is simple, emarginate, and its corners dentate; the lateral lobes are several times dichotomously incised, the ultimate subdivisions emarginate or bidentate; frequently the middle lobes are once more dichotomous than the basal lobes, so as to have twice the number of incisions. The lobes and their subdivisions are alike cuneate, approximate and radiant.

In the second section the fronds are smaller, and elliptic rather than circular; the end lobe is somewhat exserted and diverges from the lateral lobes. The division on each side into basal and middle lobes is less marked, so that in some species the segments may more appropriately be called three- than five-lobed. As in the first section the lobes are cuneate and radiant, they are also dichotomously incised, and their ultimate subdivisions are dentate, but fewer in number than in the former, in which there are four to eight in each lateral lobĕ, whilst in this section the ultimate subdivisions in the basal and middle lobes are rarely more than two.

The third section agrees with the first, inasmuch as its species are also circular, though considerably smaller; its essential difference depends on the great breadth of the end lobes and the very slight incisions between the lateral ones.

In the fourth section there is but a single species. It is oblong, its segments are five-lobed, and their lobes slightly emarginate.

The form of the frond of the species belonging to the fifth section differs from that of the preceding ones. It is not only smaller, but each segment is again constricted, and the direction of the lobes is horizontal. Unlike the lobes in the foregoing sections, these are broadest at the base and attenuated upwards, and are of course divergent; their extremities are simply bidentate. The margin of the end is convex or straight, not emarginate.

For the reception of one species belonging to this section, Mr. Hassall has constituted his new genus *Holocystis*; but, allowing that the plants I have placed here differ in form from the other species in the genus, and giving no opinion on the propriety of their removal, I think it better at present to retain them in *Micrasterias,* because some of the forms have been regarded by Ehrenberg and Meneghini as merely young states of other species; and Kützing, who has described them as distinct, has not separated them from the other species of *Micrasterias*.

* *Frond circular; segments five-lobed; lobes approximate, the end lobe narrow.*

1. *M. denticulata* (Bréb.); frond orbicular, smooth; segments five-lobed; lobes dichotomously divided, the ultimate subdivisions truncato-emarginate with rounded angles.

Micrasterias denticulata, Brébisson, *Alg. Fal.* p. 54. t. 8. (1835).
Euastrum Rota, Ehrenberg, *Infus.* t. 12. f. 1. *a.* (1838). Bailey, *Amer. Bacil. in Amer. Journal of Science and Arts*, v. 41. t. 1. f. 22.
Micrasterias Rota, Meneghini, *Syn. Desmid. in Linnæa* 1840, p. 215.
Micrasterias rotata, Ralfs, *Annals of Nat. Hist.* v. 14. t. 6. f. 1. (1844); *Trans. of Bot. Society of Edinburgh*, v. 2. t. 10.

Common. Wales; Penzance, &c., *J. R.* Barmouth, *Rev. T. Salwey.* Near Southampton; Sussex and Kent, *Mr. Jenner.* Henfield, *Mr. Borrer.* Stevenston, Ayrshire, *Rev. D. Landsborough.* Yate near Bristol, *Mr. Broome.* Ambleside, Westmoreland, *Mr. Sidebotham.* Aberdeenshire, *Dr. Dickie* and *Mr. P. Grant.*

Falaise, *Brébisson.* Germany, *Ehrenberg.* West Point, New York, *Bailey.*

The frond is large and circular; each segment is five-lobed. The end lobe is the narrowest, is simply emarginate, and often appears more turgid than the others. The lateral lobes are dichotomously incised, their ultimate subdivisions truncato-emarginate. The lobes and subdivisions are alike cuneate, approximate, and radiating from the centre. The endochrome is bright green, and the starch vesicles scattered and conspicuous. Frequently the margin of the frond is colourless.

I once gathered sporangia of this species rather plentifully in a boggy pool near Dolgelley. The lenticular shape of the fronds and their conjugation in a crossed position render it almost impossible to obtain a satisfactory view without displacing them. The process appears similar to what takes place in *Staurastrum dejectum*; the contents of both fronds unite and form a globular sporangium between them, the connexion between this sporangium and the empty segments being ascertained merely by their contiguity and by their relative position, which undergoes no change during the movements of the sporangium. The sporangium is inclosed in a fine membrane, and its surface is gradually furnished with scattered, stout, elongated spines, which are at first simple and their apex obtuse, but this afterwards becomes forked or trifid, and its divisions finally are again branched and frequently more or less recurved. The sporangia are of considerable size,—a necessary consequence of the union of the contents of both fronds.

The truncate ends of the subdivisions distinguish this species from *Micrasterias rotata*, to which it is very closely allied.

Length of frond $\frac{1}{113}$ of an inch; breadth of segment $\frac{1}{138}$; breadth at constriction $\frac{1}{277}$; diameter of sporangium from $\frac{1}{357}$ to $\frac{1}{318}$; length of spine from $\frac{1}{769}$ to $\frac{1}{741}$.

Tab. VII. fig. 1. *a.* mature frond; *b.* dividing frond; *c, d, e* and *f.* different stages of sporangia; *g.* sporangium magnified 400 times.

2. *M. rotata* (Grev.); frond orbicular, smooth; segments five-lobed; lobes dichotomously incised, ultimate subdivisions bidentate.

Echinella rotata, Greville, *in Hooker's Brit. Fl.* v. 2. p. 398 (1830).
Euastrum Rota, Ehr. *Abh. der Berl. Ak.* (1831), p. 82; *Infus.* t. 12. f. 1. *c, e.*
 Kützing, *Phycologia Germanica*, p. 134.
Eutomia rotata, Harvey, *Manual of Brit. Algæ*, p. 187 (1841).
Micrasterias rotata, Ralfs, *Annals of Nat. Hist.* v. 14. p. 259. t. 6. f. 1 (1844); *Trans. of Bot. Soc. of Edinburgh*, v. 2. p. 131. t. 10.

Common. Appin, *Capt. Carmichael.* North and South Wales; Dartmoor and Penzance, *J. R.* Sussex, Surrey and Kent, *Mr. Jenner.* Near Bristol, *Mr. Broome.* Aberdeen, *Dr. Dickie.* Ambleside, *Mr. Sidebotham.*

Germany, *Ehrenberg.* Falaise, *Brébisson.* Rhode Island, *Bailey.*

M. rotata is very similar to *M. denticulata*, but differs from it in having the ultimate subdivisions dentated. The end lobe is usually slightly exserted, and the middle lobes have eight subdivisions instead of four; I have nevertheless some doubt whether this plant is not a variety of *M. denticulata*, as the angles are sometimes merely acute instead of being prolonged into teeth.

Length of frond $\frac{1}{91}$ of an inch; breadth $\frac{1}{104}$.

Tab. VIII. fig. 1. *a.* frond with endochrome; *b.* empty frond.

3. *M. fimbriata*, ——; frond orbicular, smooth; segments five-lobed; lobes dichotomously incised; ultimate subdivisions obtusely emarginate, spinoso-mucronate.

Micrasterias Rota, in part, Ehr. *Infus.* t. 12. f. *d.* (1838).

In a boggy pool near Dolgelley, *J. R.*

Germany, *Ehrenberg.*

The frond is large and circular; the segments five-lobed; the end lobe linear-cuneate, having a broad shallow notch and two or three spines or mucros at each angle, the angles rounded. As in the preceding species, the lateral lobes are dichotomously incised, but their incisions are more shallow, the ultimate subdivisions rounded and slightly emarginate, and each furnished with two spines, which are often curved.

The basal lobes are twice and the middle lobes usually three times dichotomous, consequently the latter may have eight subdivisions whilst the former have only four. The end lobe is somewhat exserted, and frequently also there is an elongation of the subdivisions of the basal lobes which border the central constriction.

Length of frond $\frac{1}{108}$ of an inch; breadth $\frac{1}{119}$.

Tab. VIII. fig. 2. *a.* empty frond; *b.* frond with endochrome.

4. **M. radiosa** (Ag.); frond orbicular, smooth; segments five-lobed; lobes dichotomously divided; ultimate subdivisions inflated, attenuated at the end.

Micrasterias radiosa, Ag. Bot. Zeit. 1827.
Euastrum Sol, Ehr. Verbreitung und Einfluss des mikroskopischen Lebens in Süd- und Nord-Amerika, 1843, t. 4. f. 16.

In a small pool a little below the outlet of Llyn Gwernan near Dolgelley, *J. R.*

Maine to Virginia, *Bailey*.

The frond, which is about the size of that of *M. rotata*, is orbicular; segments five-lobed. The end lobe is narrow-cuneate, emarginate, and its angles dentate. The lateral lobes are dichotomously divided as in all the preceding species, but the subdivisions, especially the ultimate ones, are somewhat inflated, and their extremities are either bidentate or taper into a short mucro.

Micrasterias radiosa approaches *M. rotata* more nearly than any other species, but it differs from all in its inflated subdivisions. The extremities of the subdivisions in all the preceding species are truncate, but in this they usually taper into a point.

Length of frond $\frac{1}{138}$ of an inch; breadth $\frac{1}{138}$.
Tab. VIII. fig. 3. *a*. frond with endochrome; *b*. empty frond.

5. **M. papillifera** (Bréb.); frond orbicular, with marginal gland-like teeth; segments five-lobed; lobes dichotomously incised; the principal sinuses bordered by a row of minute granules.

Micrasterias papillifera, Brébisson *in lit. cum icone* (1846).

Hab. Dolgelley and Penzance, *J. R.* Sussex, Hants and Surrey, *Mr. Jenner*. Ambleside, *Mr. Sidebotham*. Aberdeen, *Dr. Dickie* and *Mr. P. Grant*.

Falaise, *Brébisson*. West Point, New York, *Bailey*.

The frond is smaller than in *M. rotata*; the segments five-lobed; the end lobe is nearly as broad as the others and emarginate, its angles are dentate. The lateral lobes are usually equal and dichotomously divided, the incisions rather shallow, and the ultimate subdivisions terminated by two or three gland-like teeth.

The surface of the frond has a row of minute hyaline granules or spines bordering the principal sinuses, but in other respects it is smooth. Except in the empty frond, these granules, though sometimes evident, are frequently detected with difficulty; in order to perceive them, the lens should be gradually withdrawn until the endochrome becomes less distinct, when they will appear like puncta.

The endochrome is brownish green, and the starch vesicles minute and numerous.

The sporangia, which I have gathered at Dolgelley, are similar to those of *M. denticulata*, but smaller.

M. papillifera is a very distinct species, and its characters were detected about the same time by Mr. Jenner and myself. It may be distinguished at first sight from all the preceding species by its smaller size, the gland-like appearance of the marginal teeth, the browner colour of the endochrome, and the smaller and more crowded starch vesicles. It is more difficult to perceive the puncta-like granules which border the sinuses.

This plant agrees more nearly with *M. apiculata*, a species not yet detected in Britain; but in *M. apiculata* the endochrome is of a brighter green, and the spines, which are scattered over the surface, are far more conspicuous.

Length of frond from $\frac{1}{221}$ to $\frac{1}{205}$ of an inch; breadth from $\frac{1}{238}$ to $\frac{1}{211}$; diameter of sporangium $\frac{1}{568}$; length of spine $\frac{1}{1562}$.

Tab. IX. fig. 1. *a.* frond with endochrome; *b, c.* empty fronds; *d* and *e.* end and lateral views from drawings by M. De Brébisson; *f.* sporangium; *g.* sporangium magnified 400 times.

** *Frond subelliptic; segments three- or five-lobed; lobes radiant, the end one somewhat exserted and divergent.*

6. **M. furcata** (Ag.); segments five-lobed; lobes bifid, their divisions linear, divergent, and forked at the apex.

Micrasterias furcata, Ag. *Bot. Zeit.* (1827). Kütz. *Synop. Diatom. in Linnæa* 1833, p. 603. Menegh. *Synop. Desmid. in Linnæa* 1840, p. 216.
Micrasterias Melitensis, Ralfs, *in Annals of Nat. Hist.* v. 14. p. 260. t. 6. f. 2. (bad) (1844); *Trans. of Bot. Soc. of Edinburgh*, p. 132.
Micrasterias radiata, Hass. *Brit. Freshwater Alg.* p. 386 (1845).

Hab. In a small pool a little below the outlet of Llyn Gwernan near Dolgelley, also in the stream that issues from the lake: very rare, *J. R.*

Worden's Pond, Rhode Island, *Bailey*.

The frond is smaller than that of *M. rotata*; its outline is broadly elliptic, sometimes almost circular. Each segment is deeply five-lobed; all the lobes are bifid and their subdivisions linear and forked at the extremity, the forks being usually incurved. The end lobe is broadest and somewhat exserted; its divisions are more divergent and the sinus more rounded than in the lateral lobes. The endochrome is green with scattered starch vesicles, and seldom extends beyond the bifurcation of the lobes.

M. furcata bears some resemblance to *M. Crux-Melitensis*, but is far more elegant, its incisions are deeper, its divisions more slender, and the terminal notch is incurved in a forceps-like manner.

Length of frond $\frac{1}{135}$ of an inch; breadth $\frac{1}{156}$.

Tab. IX. fig. 2. *a.* frond with endochrome; *b.* empty frond.

7. **M. Crux-Melitensis** (Ehr.); frond rotundato-elliptic; segments sub-five-lobed; lobes bifid, subdivisions short, stout and bidentate at the apex.

Euastrum Crux-Melitensis, Ehr. *Abh. d. Berl. Ak.* p. 82 (1831); *Infus.* p. 162. t. 12. f. 3. *Pritch. Infus.* p. 196. f. 124. Bailey, *Amer. Bacil.* t. 1. f. 23.

Micrasterias melitensis, Menegh. *Synop. Desmid. in Linnæa* 1840, p. 216.

Hab. Pools, very rare. Dolgelley, *J. R.* Congleton, Cheshire, and Ambleside, Westmoreland, *Mr. Sidebotham.* Henfield, *Mr. Jenner.*

Germany, *Ehrenberg.* Falaise, *Brébisson.* Maine to Virginia, *Bailey.*

The frond is about the size of that of *M. furcata* and similarly divided, but the incisions are less deep, the subdivisions stouter and less divergent, and their extremities are bidentate rather than forked; the end lobe also is more suddenly contracted beneath the bifurcation. The colouring matter extends nearly to the margin.

Mr. Jenner finds the empty fronds very delicately punctated.

Length of frond $\frac{1}{206}$ of an inch; breadth $\frac{1}{221}$.

Tab. IX. fig. 3. *a.* frond with endochrome; *b.* empty frond.

8. *M. morsa* ——; frond angular-elliptic; segments three-lobed, end lobe with bipartite angles; lateral lobes broad, margin concave inciso-serrate.

Euastrum (No. 4), Bailey, *Amer. Bacil.* t. 1. f. 25.

α. Serratures distinct.

β. Serratures obscure.

Hab. α. In a bog near Llyn Gwernan, Dolgelley, *J. R.*
β. Pools near the Quaker's Chapel, Dolgelley, *J. R.* Ashdown Forest, *Mr. Jenner.*

New York to Virginia, *Bailey.*

The frond of *Micrasterias morsa* is as large as that of *M. Crux-Melitensis.* The segments are three-lobed; the end lobe is broad, cuneate and somewhat exserted, has a broad shallow notch, and at each angle is bifid; the divisions are narrow, and each terminated by two or three minute teeth. The lateral lobes are broad and cuneate, and their margin is concave, symmetrically incised and serrated.

In β. the characters are less strongly marked, the lateral lobes are waved rather than toothed, and also more irregular.

Micrasterias morsa differs from *M. Crux-Melitensis* in the bifid angles of the end lobes and in the concave and serrated margin, as well as the less-divided state of the lateral lobes. The concave margin of the lateral lobes produces a more angular appearance of the frond than in the other species, and the plant, especially in the variety β, acquires some similarity of outline to *Euastrum verrucosum*, but its serrated margin and the bifid angles of its end lobes oblige me to place it in *Micrasterias.*

Length of frond $\frac{1}{204}$ of an inch; breadth $\frac{1}{254}$.

Tab. X. fig. 1. *a.* frond with endochrome; *b, c.* empty fronds; *d, e.* fronds of β.

*** *Fronds orbicular; segments obscurely five-lobed, the end one broadest.*

9. **M. truncata** (Corda); frond orbicular; segments with five shallow lobes, the end one very broad, truncated, lateral ones inciso-dentate.

Cosmarium truncatum, Corda, *Alm. de Carlsb.* 1835, p. 121. f. 23.
Euastrum Rota, Ehr. *Infus.* t. 12. f. 1. *g, h.* (1838).
Micrasterias Rota (young), Menegh. *Synop. Desmid.* p. 215 (1840).
Euastrum semiradiatum, Bréb. *in Menegh. Synop. Desmid.* (1840). Kützing, *Phycologia Germ.* p. 134.
Euastrum (No. 4), Bailey, *Amer. Bacil.* f. 24. (1841).
Micrasterias rotata, Ralfs, *Annals of Nat. Hist.* v. 14. t. 6. f. 1. *b.* (inferior) (1844); *Trans. of Bot. Soc. of Edin.* v. 2. t. 10.
Micrasterias truncata, Bréb. *in lit.* (1846).

Common. Sussex and Hampshire, *Mr. Jenner.* Dolgelley, Penzance, &c., *J. R.* Aberdeen, *Dr. Dickie* and *Mr. P. Grant.* Near Bristol, *Mr. Broome.* Ambleside, Westmoreland, *Mr. Sidebotham.* Kerry, *Mr. Andrews.*

Germany, *Corda, Kützing.* Falaise, *Brébisson.* United States, *Bailey.*

The frond is minute and circular, the segments are five-lobed, and the separating sinuses shallow, especially those between the lateral and intermediate lobes. The corners of the terminal lobe, which is broadly cuneate and truncate, are bidentate, and the lateral lobes are inciso-dentate.

Mr. Jenner and myself have hitherto differed respecting this plant. He has considered it distinct, whilst I have been disposed to regard it as the young state of *M. rotata,* and it was figured as such in the plate (already engraved) of that species; but I have learnt since from M. de Brébisson and Professor Kützing, that they entertained no doubt of the distinctness of this species and had so described it, and I have myself met with specimens containing in abundance starch vesicles; I therefore hasten to admit its claims to the rank of a distinct species.

Length of frond $\frac{1}{240}$ of an inch; breadth $\frac{1}{250}$.

Tab. VIII. fig. 4. and Tab. X. fig. 5. *a.* frond with endochrome; *b.* empty frond; *c.* end view; *d.* side view.

10. **M. crenata** (Bréb.); frond orbicular; segments with five shallow lobes, the end one very broad and convex at the margin, lateral ones nearly entire.

Micrasterias rotata, Ralfs, *Ann. of Nat. Hist.* v. 14. t. 6. f. 1. *b.* (superior) (1844); *Trans. of Bot. Soc. of Edin.* v. 2. t. 10.
Micrasterias crenata, Bréb. *in lit.* (1846).

Dolgelley, *J. R.* Sussex, *Mr. Jenner.*

The frond is minute, of the same size as that of *M. truncata,* and circular. The segments are five-lobed, the terminal one broad, cuneate, entire and rounded at the margin. The lateral lobes are sometimes entire, but more usually crenate, occasionally having a few inconspicuous teeth.

M. crenata differs from *M. truncata* in the convex margin of its end lobes and the absence of teeth on the lateral ones, but I have seen some specimens which make it doubtful whether they be really distinct.

One figure of it was engraved as the young state of *M. denticulata*, before reasons nearly similar to those which I have stated under the foregoing species had induced me to describe it as distinct.

Length of frond $\frac{1}{244}$ of an inch; breadth $\frac{1}{263}$.

Tab. VII. fig. 2. and Tab. X. fig. 4. *a.* frond with endochrome; *b.* empty frond.

**** *Fronds oblong.*

11. *M. Jenneri*; frond oblong, minutely granulated; segments five-lobed, lobes closely approximate, cuneate, lateral ones obscurely bipartite, the subdivisions emarginate.

a. Granules appearing like mere puncta.

β. Granules larger, giving a dentate appearance to the margin.

a. Greatham Bogs, Fittleworth, and Ashdown Forest, Sussex, and near Southampton, *Mr. Jenner.* Dolgelley, *J. R.*

β. Medhurst, and Ashdown Forest, Sussex, *Mr. Jenner.*

Fronds large, twice as long as broad, oblong or quadrilateral. Segments five-lobed; lobes cuneate, approximate, lateral ones slightly bipartite and the subdivisions truncato-emarginate. Endochrome green with scattered vesicles. The transverse view of the frond fusiform.

The surface of the frond is furnished with minute pearly granules which usually look like mere puncta, but in some specimens gathered by Mr. Jenner in Sussex they are larger and give the margin a dentate appearance.

This puzzling plant almost seems to unite *Micrasterias* with *Euastrum*. It agrees with the latter genus in figure, and the lobes also at first sight appear more like those of a species of *Euastrum* than one of *Micrasterias*; but they have incisions, although inconspicuous, which divide them into two portions, each slightly emarginate. A transverse view shows the absence of the inflated protuberances always found in true species of *Euastrum*.

Micrasterias Jenneri differs from all other species of *Micrasterias* in its oblong fronds. It may easily be distinguished from *Euastrum verrucosum* and *E. crassum* by its five-lobed segments, and from *E. oblongum* not only by its more quadrangular form, but by the lobes being so closely in contact as almost to conceal the separating sinuses.

Length of frond $\frac{1}{147}$ of an inch; breadth $\frac{1}{209}$.

Tab. XI. fig. 1. *a.* frond with endochrome; *b.* empty frond; *c.* frond magnified 400 times; *d.* transverse view.

***** *Lobes horizontal, attenuated, bidentate.*

12. *M. oscitans* (Ralfs); frond with convex ends; segments constricted; lobes horizontal, conical, bidentate.

Micrasterias oscitans, Ralfs, *in Jenner's Fl. of Tunbridge Wells*, p. 198 (1845).

Holocystis oscitans, Hassall, *Brit. Freshwater Algæ*, p. 386. t. 90. f. 4. (1845).

Dolgelley, *J. R.* Fittleworth near Petworth, Midhurst, and Ashdown Forest, Sussex, *Mr. Jenner*. Ambleside, *Mr. Sidebotham*. Aberdeen, *Mr. P. Grant*.

West Point, New York, *Bailey*.

Frond smooth, nearly as large as that of *Micrasterias Crux-Melitensis*, and convex at its ends. Segments deeply constricted; lobes horizontal, conical and bidentate at the apex, the basal ones longest.

The transverse view is fusiform, with two teeth at each end and a small circular central opening at the connexion of the segments. Endochrome green with scattered vesicles. The empty frond is minutely punctate.

From the resemblance of this plant to Ehrenberg's figure of young *M. Crux-Melitensis*, I at first considered it the immature state of another species; but having repeatedly gathered it unmixed with any species of *Micrasterias* to which it could belong, and watched it carefully for some time without detecting any alteration, I am obliged to consider it as distinct. Its size alone would forbid the supposition that it belonged to *M. Crux-Melitensis*; it is thicker than any other species of *Micrasterias*, and finally its starch vesicles are too abundant for a young plant. Subsequently Mr. Jenner gathered it in Sussex, and he fully agrees with my opinion that it is a distinct species.

Mr. Hassall, who separates it from *Micrasterias*, has overlooked the bidentate extremities of the lobes, notwithstanding their presence in the Dolgelley specimens, from which his figure was taken.

Length of frond $\frac{1}{56}$ of an inch; breadth of basal lobe $\frac{1}{211}$; breadth of end lobe $\frac{1}{269}$.

Tab. X. fig. 2. *a*. frond with endochrome; *b*. empty frond; *c*. transverse view.

13. **M. pinnatifida** (Kütz.); frond plane, its ends straight; segments deeply constricted; lobes horizontal, triangular, bidentate.

Euastrum (No. 7), Bailey, *Amer. Bacill.* t. 1. f. 29. (1841).
Euastrum pinnatifidum, Kütz. *Phycologia Germ.* p. 134. (1845).

Hab. Dolgelley, *J. R.* Ambleside, *Mr. Sidebotham*.
United States, *Bailey*. Germany, *Kützing*. Falaise, *Brébisson*.

The frond is very minute and plane; its segments are deeply constricted, and straight or slightly concave at the end. The lobes are horizontal, triangular and bidentate at the apex, the basal ones longest.

M. pinnatifida is a much smaller species than *M. oscitans*, but resembles it in form. The surface however is flatter, the end margin is straight or slightly concave instead of convex, and the lobes are more tapering so as to appear triangular; the colour also is paler.

I have not detected starch vesicles in this plant, and might have taken it for the young state of *M. Crux-Melitensis*, but its form is so similar to that of *M. oscitans*, that the claim of the latter undoubted species to be considered distinct would be invalidated if *M. pinnatifida* should be proved an immature state of *M. Crux-Melitensis*.

Length of frond $\frac{1}{440}$ of an inch; breadth at basal lobes $\frac{1}{392}$; breadth at end lobes $\frac{1}{555}$; breadth at constriction between the lobes $\frac{1}{1818}$.

Tab. X. fig. 3. *a, b.* fronds with endochrome.

7. EUASTRUM, *Ehr.*

Frond simple, compressed, deeply divided into two segments which are emarginate at their ends, lobed or sinuated, generally pyramidal and furnished with circular inflations.

The fronds are simple, longer than broad, often oblong, compressed, and so deeply constricted that their segments seem only united by a narrow chord. The generally pyramidal segments are broadest at their bases, and are there in such close apposition for their entire breadth as nearly to conceal the notch on each side until the endochrome has collapsed. They are attenuated towards the ends, which in the adult state are almost always more or less emarginate, and their sides are more or less lobed or sinuated. The surface is irregular with inflated prominences, which also form tubercle-like projections along the margins; their number and situation are, probably, constant in the adult fronds of the same species and different in distinct species. A transverse view is (when the segments are separated) the best method of ascertaining their number; the terminal lobe has similar prominences.

In *Euastrum*, Ehrenberg includes *Micrasterias*, Ag. (not *Micrasterias*, Ehr.) and *Cosmarium*, and in this he is followed by Kützing in his 'Phycologia Germanica.' Meneghini separates *Micrasterias* from *Euastrum*, but includes the latter in *Cosmarium*. *Euastrum* appears to me distinct from both, and especially from *Cosmarium*.

Euastrum agrees with *Micrasterias* in having lobes and emarginate ends, but the lobes are not incised, nor do they radiate from the centre, and the inflated projections will distinguish it not only from *Micrasterias*, but from every other genus in the family. From *Cosmarium* it differs also in the lobed and emarginate segments.

I have divided this genus into three sections. In the first section the fronds are comparatively large, and appear to the naked eye like roundish or oblong dots. The segments are distinctly lobed; the terminal lobe, cuneate and itself emarginate, is partly included in a notch between the projections of the lateral lobes, and the sinuses which separate it from them are deep, and directed inwards and downwards.

In the second section the fronds are more minute and scarcely visible to the naked eye; the segments are less decidedly lobed, but the margin is crenate or sinuated, and the terminal portion unites with the basal by a neck-like contraction of the segment, and is therefore never included within a notch; the corners are rounded. The outline of a segment has some resemblance to that of a decanter or water-bottle.

The third section contains those species which do not well agree with those in the preceding ones. The fronds are extremely minute, the segments are generally still less lobed than in the last, and the form of the front view is more irregular, and frequently differs in having a process or an acute angle at either the corners or the sides of the terminal portion.

Several species have been observed in a conjugated state.

* *Segments of the frond deeply lobed; the terminal lobe cuneate, and partly included in a notch formed by the projection of the lateral lobes.*

1. *E. verrucosum* (Ehr.); frond rough with conic granules; the segments three-lobed; lobes broadly cuneate with a broad shallow notch.

Euastrum verrucosum, Ehr. *Abh. d. Berl. Ak.* p. 247 (1833); *Infus.* p. 162. t. 12. f. 5. Ralfs, *Annals of Nat. Hist.* v. 14. p. 189. t. 6. f. 3; *Trans. of Bot. Soc. of Edin.* v. 2. p. 125. t. 10. Hass. *Brit. Alg.* p. 379. Kütz. *Phycologia Germanica*, p. 135.
Cosmarium verrucosum, Menegh. *Synop. Desmid. in Linnæa* 1844, p. 222.

In pools. Cheshunt, *Mr. Hassall.* Weston Bogs near Southampton; Rusthall Common and elsewhere near Tunbridge Wells; Ashdown Forest, Sussex; Hampshire; and Reigate, Surrey, *Mr. Jenner.* Penzance and Dolgelley, *J. R.* Near Ambleside, Westmoreland, *Mr. Sidebotham.* Aberdeenshire and Banffshire, *Mr. P. Grant.*

Germany, *Ehrenberg, Kützing.* Falaise, *Brébisson.* New York, *Bailey.*

Frond rather longer than broad, angular; the segments, which slightly diverge from each other, are deeply three-lobed; the lobes are broad, cuneate, with a broad shallow notch. Surface of the frond furnished with numerous conic granules, which give the margins a dentated appearance, especially the more prominent parts: each segment has two circular prominences near the base; on these the granules form two or three concentric circles with a granule in the centre; the terminal lobe has two similar but smaller prominences.

The side view, which is not so broad as the front one, is inflated at the base, attenuated upwards into a short neck, and emarginate and slightly dilated at the end. A transverse view is oblong, with three inflations at each side and a smaller one at each end. The terminal lobe, as seen by an end view, is four-lobed.

Euastrum verrucosum, when once seen, is not liable to be confounded with any other species, but may be known by the conic granules giving a dentate appearance to the outline.

Length of frond $\frac{1}{267}$ of an inch; breadth $\frac{1}{270}$.

Tab. XI. fig. 2. *a.* frond with endochrome; *b.* empty frond; *c.* side view; *d.* end view.

2. **E. oblongum** (Grev.); frond smooth, oblong; segments five-lobed; lobes cuneate, emarginate, the terminal one partly included between the lateral ones.

Echinella oblonga, Grev. *in Hook. Br. Fl.* v. 2. p. 398 (1830).
Euastrum Pecten, Ehr. *Abh. d. Berl. Ak.* 1831, p. 82; *Infus.* p. 162. t. 12. f. 4. Kütz. *Phy. Germ.* p. 135.
Cosmarium sinuosum, Corda, *Alm. de Carlsb.* 1835, p. 121. t. 2. f. 21.
Micrasterias sinuata, Bréb. *Alg. Fal.* p. 55. t. 7. (1835).
Cosmarium oblongum, Bréb. *in Menegh. Synop. Desmid. in Linnæa* 1840, p. 221.
Eutomia oblonga, Harv. *Br. Alg.* p. 188 (1841).
Euastrum oblongum, Ralfs, *in Annals of Nat. Hist.* v. 14. p. 189. t. 6. f. 4. (1844); *Trans. Bot. Soc. Edin.* v. 2. p. 126. t. 10. Hass. *Brit. Freshwater Alg.* p. 380.

Hab. Pools; common. Appin, *Capt. Carmichael.* Penzance; Dolgelley and Carnarvon, *J. R.* Warbleton, Henfield, &c., Sussex; Tunbridge Wells; Reigate, Surrey; and Weston Bog near Southampton, *Mr. Jenner.* Brookhouse Moss, near Congleton, Cheshire, and near Ambleside, Westmoreland, *Mr. Sidebotham.* Aberdeen, *Mr. P. Grant* and *Dr. Dickie.*

Falaise, *Brébisson.* Germany, *Ehrenberg, Kützing.* New York; Rhode Island and Virginia, *Bailey.*

Frond comparatively large, appearing to the naked eye like a small dot, oblong, three or four times longer than broad; each segment divided into five lobes in a pinnatifid manner. The lateral lobes are broad, cuneate, with a broad shallow notch. The terminal lobe is cuneate and its notch linear; the corners of all the lobes are rounded. The empty fronds are punctated.

The transverse view is three times longer than broad, and has three rather distant inflations or lobes on each side (the largest in the centre) and one at each end. I have not seen the end view myself, but Mr. Jenner finds the terminal lobe twice as long as broad, slightly constricted at the middle, its ends notched rather than lobed.

The sporangia are orbicular, and furnished with numerous conical tubercles; they have been met with only near Dolgelley.

The young fronds are smaller and have the middle lobe entire.

Length of frond $\frac{1}{156}$ of an inch; breadth $\frac{1}{282}$; breadth of end lobe $\frac{1}{555}$; diameter of sporangium $\frac{1}{175}$; length of tubercle $\frac{1}{1000}$.

Tab. XII. *a.* mature frond with endochrome; *b.* empty frond; *c.* mature frond of small variety; *d.* transverse view; *e.* transverse view of small variety; *f.* end view; *g.* sporangium.

3. *E. crassum* (Bréb.); frond smooth; segments three-lobed, subquadrilateral; terminal lobe cuneate, partly included in a notch formed by the lateral lobes.

Cosmarium crassum, Bréb. *in Menegh. Synop. Desm.* p. 222. (1840).
Euastrum Pelta, Ralfs, *in Annals of Nat. Hist.* v. 14. p. 190. t. 7. f. 1. (1844), (not *Cosm. Pelta*, Corda, according to Brébisson); *Trans. Bot. Soc. of Edin.* p. 126. Hass. *Br. Freshwater Algæ*, p. 380.
Euastrum crassum, Kütz. *Phycologia Germ.* p. 135 (1845).
Euastrum (No. 5), Bailey, *Amer. Bacil.* t. 1. f. 26.

β. Smaller, sides more concave.

Hab. Pools; common. Dolgelley, Penzance, *J. R.* Weston Bogs near Southampton; Ashdown Forest, Sussex, and Fisher's Castle, Kent, *Mr. Jenner.* Ambleside, Westmoreland, *Mr. Sidebotham.* Aberdeen, *Mr. P. Grant* and *Dr. Dickie.* Ben Muich Dhu and Glen Lin, Aberdeenshire, *Dr. Dickie.*

β. Dolgelley, *J. R.*
Falaise, *Brébisson.* United States, *Bailey.* Germany, *Kützing.*

Frond comparatively large, visible to the naked eye, about three times longer than broad; of a quadrilateral form, with rounded ends; terminal lobe cuneate, partly included between the ends of the lateral lobes, rounded and emarginate, the notch closed; the segments of the frond are broad, three-lobed, or rather each segment has a subquadrate base and a terminal lobe. The basal portion is not attenuated, and each lateral margin has a broad shallow sinus, in which there is frequently a slight intermediate rounded projection.

The transverse view is two or three times longer than broad, with three lobes on each side and one at each end.

The empty frond is punctated.

The var. β, regarded by Brébisson as a distinct species, is smaller, the segments have more concave sides, their base is more inflated, and the angles which include the terminal lobe are more elongated.

The *E. cornutum*, Kütz., differs in having the outer angles of the lateral lobes elongated into processes.

Length of frond from $\frac{1}{193}$ to $\frac{1}{132}$ of an inch; breadth from $\frac{1}{263}$ to $\frac{1}{260}$; breadth of end lobe $\frac{1}{537}$.

Tab. XI. fig. 3. *a.* frond with endochrome; *b.* empty frond; *c.* transverse view; *d.* end view; *e.* frond of β; *f.* end view.

** *Segments sinuated; terminal lobe exserted and united with the basal portion by a distinct neck.*

4. *E. pinnatum*, ——; segments five-lobed; end lobe exserted, dilated; upper margin of lobes horizontal.

Dolgelley, rare, *J. R.*

Frond oblong, and rather smaller than that of *E. oblongum*. Segments deeply five-lobed in a pinnatifid manner; the terminal lobe is exserted, dilated and emarginate; the basal lobes are emarginate and the intermediate ones smaller and entire; the upper margins of all are horizontal. The transverse view is oblong, with three lobes or projections on each side and one at each end. The empty frond is punctate, and the inflated projections are strongly marked.

Euastrum pinnatum somewhat resembles, but is smaller than *E. oblongum*; the terminal lobe joins the others with a distinct neck; the intermediate lobes are always simple and smaller than the basal, and all the sinuses are more rounded. There is more difficulty in separating *Euastrum affine*; still, when it is compared with *E. pinnatum* the distinction is easily perceived, for the latter is a larger plant, and is also distinctly five-lobed; but in the former the intermediate lobes resemble tubercles, as the sinuses between them and the basal ones are much shallower.

Length of frond $\frac{1}{188}$ of an inch; breadth at basal lobes $\frac{1}{357}$; breadth at middle lobes $\frac{1}{454}$; breadth at end $\frac{1}{625}$; breadth of neck $\frac{1}{1101}$.

Tab. XIII. fig. 1. *a.* frond with endochrome; *b.* empty frond; *c.* side view; *d.* transverse view; *e.* end view.

5. **E. humerosum,** ——— ; **segments with terminal lobe dilated, emarginate; neck partly included between the elongated middle lobes, which resemble processes; basal lobes emarginate.**

Dolgelley, *J. R.* New Forest, Hants, *Mr. Jenner.*

Frond smooth, two or three times longer than broad; segments sub-five-lobed; the basal lobes are emarginate; the terminal one is dilated, its notch linear, and the neck is short and partly included between the elongated processes or tubercles which represent the middle lobes. The transverse view has three lobes on each side and one at each end. The empty frond is minutely dotted.

Euastrum humerosum corresponds in size with *E. affine*, of which I formerly considered it a variety, but Mr. Jenner finds that they differ essentially in the transverse view. The present species is distinguished from all in this section by its included neck.

Length of frond $\frac{1}{225}$ of an inch; breadth of segment at base $\frac{1}{382}$; breadth at end $\frac{1}{727}$; breadth of neck $\frac{1}{1199}$; length of projections at side of neck $\frac{1}{2314}$.

Tab. XIII. fig. 2. *a.* frond with endochrome; *b.* empty frond; *c.* transverse view.

6. **E. affine** (Ralfs); **segments three-lobed, with intermediate tubercles; lobes emarginate, the end one dilated, its notch linear.**

Euastrum affine, Ralfs, *Annals of Nat. Hist.* v. 14. p. 191. t. 7. f. 3. (1844); *Trans. of Bot. Soc. of Edin.* v. 2. p. 128. Hass. *Brit. Alg.* p. 382.

Dolgelley, *J. R.* Ambleside, *Mr. Sidebotham.* Sussex and Hampshire, *Mr. Jenner.* Aberdeenshire, *Dr. Dickie* and *Mr. P. Grant.*

Falaise, France, *Brébisson.*

Frond about three times as long as broad; segments pyramidal, three-lobed, and having on each side a tubercle, which replaces the middle lobes; the end lobe is dilated and its notch linear; the basal lobes are emarginate and suddenly contracted beneath the tubercles, which are smaller than in the last species, and have their upper margins nearly horizontal, by which it may be distinguished from it. The transverse view also has four lobes on each side instead of three. The empty frond is punctate.

Euastrum affine differs from *E. pinnatum* by its smaller size, by being sinuated, not pinnatifid, and by the lobes being less prominent in the transverse view. In *E. pinnatum* we find the segments distinctly five-lobed, but in this and some other species the middle lobes are replaced by tubercles more or less developed.

Length of frond $\frac{1}{230}$ of an inch; breadth at basal lobe $\frac{1}{458}$; breadth at middle lobes $\frac{1}{659}$; breadth of neck $\frac{1}{1204}$; breadth at end $\frac{1}{917}$.

Tab. XIII. fig. 3. *a*. mature frond with endochrome; *b*. empty frond; *c*. side view; *d*. end view; *e*. transverse view.

7. **E. ampullaceum** ——— ; segments short, with inflated base, small intermediate tubercles, and a dilated, notched, terminal lobe.

Euastrum ———, Hass. *Brit. Freshwater Algæ*, t. 100. f. 11.

Dolgelley, *J. R.* Ambleside, Westmoreland, *Mr. Sidebotham*. Near Tunbridge Wells and near Storrington, Sussex; and Hampshire, *Mr. Jenner*.

Frond smooth, twice as long as broad, rather smaller than *Euastrum affine*. The base is much-inflated, and is separated from the small intermediate tubercles by a slight constriction. The terminal lobe is dilated, its outline usually more convex than that of the other species, and it tapers more gradually into the very short neck, hence it appears more cuneate or fan-shaped; its notch is linear.

Euastrum ampullaceum may be best recognized by its short segments, in which and its broad inflated base it differs from *E. affine*, whilst the distinctly dilated terminal lobe separates it from *E. Didelta*.

The empty frond is punctate, and the inflated protuberances on its front surface are indistinct.

Length of frond $\frac{1}{274}$ of an inch; breadth of base $\frac{1}{394}$; breadth at end $\frac{1}{788}$; breadth of neck $\frac{1}{1173}$.

Tab. XIII. fig. 4. *a*. frond with endochrome; *b*. empty frond; *c*. side view; *d*. transverse view.

8. *E. insigne* (Hass.); segments inflated at the base and tapering upwards into a slender neck; end dilated, emarginate; transverse view with two distant lobes on each side and one at each end.

Euastrum insigne, Hass. *Brit. Alg.* t. 91. f. 2. (1845).

Dolgelley, *J. R.* Midhurst, and Fittleworth Common near Petworth; and New Forest, Hants, *Mr. Jenner*. Ambleside, *Mr. Sidebotham*. Moss Hagg, Banffshire (altitude 3000 feet), *Mr. P. Grant*.

Frond rather smaller than that of *Euastrum Didelta*, about three times as long as broad; each segment is inflated at its base, the sides of which are entire, and tapers upwards into a slender neck without lateral tubercles; terminal lobe dilated, its notch linear and frequently obscured by the inflated prominences. The end view is cruciform.

The transverse view differs from that of every other species, for in them the outline is elliptical, or broadest at the centre, but in this plant the sides are straight, with a little constriction at the centre and a slight swelling or lobe near each extremity; indeed without the end lobes the form would be nearly quadrangular. The empty frond is punctate.

Mr. Sidebotham has gathered a conjugated specimen of this species near Ambleside.

The inflated base and slender neck, conjoined with its dilated end, are sufficiently characteristic.

To avoid a needless multiplication of synonyms, I have here adopted the name under which Mr. Hassall has figured it in his work on the British Freshwater Algæ, although in a paper read before the Botanical Society of Edinburgh I had previously called it *Euastrum gracile*, and had distributed specimens under that name, from some of which indeed Mr. Hassall obtained his knowledge of the species, a circumstance that he has omitted to notice.

Length of frond $\frac{1}{232}$ of an inch; breadth at base $\frac{1}{416}$; breadth at end $\frac{1}{806}$; breadth of neck $\frac{1}{1422}$.

Tab. XIII. fig. 6. *a*. frond with endochrome; *b*. empty frond; *c*. side view; *d*. transverse view; *e*. end view.

9. *E. Didelta* (Turp.); segments with inflated base, intermediate tubercles, and notched and scarcely dilated end; transverse view, four shallow lobes on each side and one at each end.

Heterocarpella Didelta, Turp. *Mem.* p. 295 (1828).
Cosmarium Didelta, Menegh. *Synop. Desmid. in Linnæa* 1840, p. 219.
Euastrum Didelta, Ralfs, *Ann. of Nat. Hist.* v. 14. p. 190 (in part), t. 7. f. 2. *a, b.* (1844); *Trans. of Bot. Soc. of Edinburgh*, v. 2. p. 127. t. 11.

Common. Carnarvon, Dolgelley, Penzance, &c., *J. R.* Near Tunbridge Wells; near Battle, Henfield, Midhurst, Sussex; and Hampshire, *Mr. Jenner.* Westmoreland, *Mr. Sidebotham.* Aberdeen, *Dr. Dickie* and *Mr. P. Grant.*

Falaise, *Brébisson.* New York, *Bailey.*

Frond comparatively large, about thrice as long as broad; segments pyramidal, inflated at the base, and tapering upwards into the intermediate tubercles; the neck is broad, and the end slightly dilated, with a terminal linear notch. The end view is bilobed, and the transverse one elliptic with four slight lobes on each side and one at each end. The empty frond is punctate.

Euastrum Didelta differs from the preceding species in the slight dilatation of the end lobes, and in the end view the two lobes are entire.

The sporangia, which I have twice found at Dolgelley, are orbicular, and have subulate spines.

Length of frond $\frac{1}{185}$ of an inch; breadth at base $\frac{1}{357}$; breadth at end $\frac{1}{752}$; breadth at constriction $\frac{1}{1250}$; thickness in side view $\frac{1}{454}$.

Tab. XIV. fig. 1. *a.* frond with endochrome; *b.* empty frond; *c.* transverse view; *d.* end view.

10. *E. ansatum* (Ehr.); segments inflated at the base, and tapering upwards to the notched but not dilated extremity; transverse view cruciform.

Euastrum ansatum, Ehr. *Infus.* p. 162. t. 12. f. 6. (1838).
Euastrum ———, Bailey, *in Amer. Journ. of Science and Arts*, v. 41. p. 295. t. 1. f. 27. (1841).
Euastrum Didelta, Ralfs, *Annals of Nat. Hist.* v. 14. p. 190 (in part). t. 7. f. 2. *c, d, e, f.* (1844); *Trans. of Bot. Soc. of Edinburgh*, v. 2. p. 127. t. 11.
Euastrum binale, Kütz. *Phy. Germ.* p. 135 (1845).

Common. North and South Wales, and Penzance, *J. R.* Hants, Surrey and Sussex, *Mr. Jenner.* Kerry, *Mr. Andrews.* Aberdeenshire and Banffshire (altitudes 200–3600 feet), *Dr. Dickie* and *Mr. Grant.* Cheshire and Westmoreland, *Mr. Sidebotham.* Near Manchester, *Mr. Williamson.* Near Bristol, *Mr. Thwaites.* Rochdale, *Mr. Coates.*

France, *Brébisson.* Germany, *Ehrenberg* and *Kützing.* United States, *Bailey.*

Frond smaller than that of *E. Didelta*, twice as long as broad; the segments inflated at the base and tapering upwards into a neck, which is not dilated at the end; the terminal notch is linear. The end view has two circular and entire lobes, and the transverse view a single inflation on each side. The empty frond is punctate.

I formerly considered this plant as the young state of *E. Didelta*, and I am not yet fully persuaded that it is distinct; but as it differs from *E. Didelta* in every view, and Mr. Jenner, whose opinion I highly value, urges their separation, I have raised this form to the rank of a species: whether it is right to consider it one, experience must decide.

Euastrum ansatum is separated from *E. circulare* by the absence of the five tubercles at the base of the segments in the front view; in respect of the other species its form is a sufficient distinction.

Length of frond $\frac{1}{315}$ of an inch; breadth at base $\frac{1}{654}$; breadth at end $\frac{1}{1298}$.

Tab. XIV. fig. 2. *a.* frond with endochrome; *b, c.* empty fronds; *d,* side view; *e.* transverse view; *f.* end view.

11. *E. circulare* (Hass.); segments three-lobed, mostly with five basal tubercles; four of them usually disposed semicircularly about the fifth; end notched, scarcely dilated.

α. Segments inflated at the base and attenuated upwards.

Euastrum circulare, Hass. *Brit. Freshwater Algæ*, p. 383. t. 90. f. 5. (1845).

β. Segments emarginate at the sides, the basal portion with five tubercles.

Euastrum sinuosum, Lenormand, *in herb.* (1845).

γ. Segments emarginate at the sides; tubercles smaller, more numerous and scattered.

α. High Beech, Essex, *Hassall*. Near Aberdeen, *Mr. P. Grant*.
β. Dolgelley, *J. R.* Ambleside, Westmoreland, *Mr. Sidebotham*. Weston Bogs near Southampton, *Mr. Jenner*.
γ. Ambleside, Westmoreland, *Mr. Sidebotham*.

δ. Falaise, *Brébisson*.

As I have seen no authentic specimens, and am not sure whether the number and position of the inflated protuberances are constant, I may have united under this name more than one species. The variety γ. especially differs from the other forms in the number and arrangement of the tubercles, but as I have seen no specimen of it, and Mr. Jenner only one or two fronds mixed with other Desmidieæ, which Mr. Sidebotham forwarded from Westmoreland, I have no means of deciding where it should be placed. At the same time I am unwilling to omit so interesting a form, and as it agrees in outline with the variety β, I have placed it here, in the hope it may be recognized.

I have gathered near Penzance a single frond, which I suppose may be the state figured by Mr. Hassall. The protuberances were distinct on one segment, but not on the other. In form it resembled *E. ansatum*; the inflations on the front surface were its most obvious distinction from that plant; in a transverse view I presume it would resemble the other varieties.

The variety β. is not uncommon near Dolgelley; the segments are emarginate at their sides and the end lobe is slightly dilated. On the front surface are five protuberances, but their arrangement is different from that in Mr. Hassall's figure; they are also less distinct and sometimes scarcely perceptible. An Euastrum, however, gathered at Falaise, for the sight of which I am indebted to Dr. Dickie, unites the β. form with protuberances exactly like those represented by Mr. Hassall.

This variety agrees in some respects with *Euastrum pectinatum* and *E. gemmatum*, but has a terminal linear notch, and the end view exhibits two nearly quadrate and slightly notched lobes.

Length of frond of var. β. $\frac{1}{325}$ of an inch; breadth at base $\frac{1}{549}$; breadth of neck $\frac{1}{1269}$; breadth at end $\frac{1}{1059}$.

Length of frond of Falaise specimen $\frac{1}{320}$ of an inch; breadth at base $\frac{1}{519}$; breadth of neck $\frac{1}{1234}$; breadth at end $\frac{1}{1149}$.

Tab. XIII. fig. 5. *a*. frond of β. with endochrome; *b*. empty frond; *d*. transverse view; *c*. empty frond of γ.

Tab. XIV. fig. 3. *a*. frond as figured by Hassall in Brit. Alg.; *b*. frond of the Falaise variety; *c*. end view.

12. *E. pectinatum* (Bréb.); segments three-lobed; terminal lobe dilated, scarcely emarginate; lateral lobes horizontal, emarginate; end view two-lobed at each end and two lobules on each side.

β. terminal lobe emarginate at each side.

Cosmarium pectinatum, Bréb. *in Menegh. Synop. Desmid. in Linnæa* 1840 p. 222.
Euastrum gemmatum, Ralfs, *Annals of Nat. Hist.* vol. 14. p. 191. t. 7. f. 4.

(1844); *Trans. of Bot. Soc. of Edinburgh,* v. 2. p. 128. t. 11. Hass. *Br. Alg.* p. 382. (not *Cosmarium gemmatum,* Bréb.).
Euastrum pectinatum, Bréb. *in lit. cum icone* (1846).

Boggy pools; not rare. Dolgelley, Penzance, *J. R.* Sussex, and Weston Bog near Southampton, *Mr. Jenner.* Brookhouse Moss, near Congleton, Cheshire; and Ambleside, Westmoreland, *Mr. Sidebotham.* Glen Derry, Aberdeenshire, *Dr. Dickie.* Aberdeen and east of the Hill of Fare, Aberdeenshire; and Moss Hagg, Banffshire, *Mr. P. Grant.*

Falaise, France, *M. De Brébisson.*

Frond two or three times longer than broad; each segment has a broad basal portion, which is somewhat quadrilateral, horizontal, and at each side emarginate; the neck is short and broad; the terminal lobe is dilated; the sides entire in a. and slightly notched in β; the terminal notch is obsolete.

A transverse view is twice as long as broad, with three lobes on each side and two at each end.

The end view shows that the terminal portion has two lobes at each end and two lobules on each side, one belonging to each lobe. The empty frond is punctate.

Euastrum pectinatum is a very distinct species, and differs from all the preceding in the absence of a terminal notch, in the lobules of the end lobe, and in having the ends emarginate in the transverse view.

I formerly considered this species as the *Cosmarium gemmatum* of Meneghini's Synopsis. I am indebted to M. De Brébisson for the information that it is distinct, and also for drawings of both plants.

Sporangia gathered at Dolgelley are orbicular, with conical tubercles or short obtuse spines.

Length of frond $\frac{1}{302}$ of an inch; breadth at base $\frac{1}{558}$; breadth at end $\frac{1}{776}$; breadth of neck $\frac{1}{1126}$; breadth at constriction $\frac{1}{2604}$; diameter of sporangium from $\frac{1}{490}$ to $\frac{1}{440}$.

Tab. XIV. fig. 5. *a.* frond with endochrome; *b.* empty frond; *c.* frond of var. β; *d.* end view; *e.* transverse view; *f.* sporangium.

13. *E. gemmatum* (Bréb.); segments three-lobed; terminal lobe dilated, broadly emarginate; basal lobes horizontal, emarginate, protuberances minutely granulate.

Cosmarium gemmatum, Bréb. Menegh. *Synop. Desmid. in Linnæa* 1840, p. 221.
Euastrum gemmatum, Kütz. *Phy. Germ.* p. 134 (1845). Bréb. *in lit. cum icone.*
Euastrum papulosum, Kütz. *Phy. Germ.* p. 135 (1845).

Bogs. Dolgelley and Penzance, *J. R.* Ambleside, *Mr. Sidebotham.* Aberdeen, *Dr. Dickie* and *Mr. P. Grant.* Weston Bogs near Southampton, and near Pulborough, Sussex, *Mr. Jenner.*

Falaise, *De Brébisson.* Nordhausen, Prussia, *Kützing.*

Frond about twice as long as broad; segments three-lobed, the basal ones horizontal, emarginate, and their projecting parts granulated. The end lobe is dilated; the neck is short, and about one-third as broad as the base; the granulated dilatations of the terminal lobe are inclined upwards, hence the end is broadly emarginate. The transverse view is oblong, with three lobes on each side and one at each end. The end view shows that the terminal lobe is cruciform.

Euastrum gemmatum is a small but elegant species, the base being quadrilateral, horizontal, and nearly three times as broad as long. It may be known from all in this section, except *E. pectinatum*, by the absence of a linear terminal notch, and from that species it may be distinguished by the smallness of the end lobe and the outward direction of its angles, and by the minute granulation of the projecting parts.

M. De Brébisson has kindly furnished me with drawings of *Euastrum gemmatum*, Bréb., and *E. papulosum*, Kütz., the latter of which he reduces to a variety of the former in a MS. list of the species of this genus; but I must confess that I know not how to make even this distinction between them.

Length of frond $\frac{1}{442}$ of an inch; breadth at base $\frac{1}{641}$; breadth at end $\frac{1}{1219}$; breadth of neck $\frac{1}{1497}$.

Tab. XIV. fig. 4. *a*. frond with endochrome; *b*. empty frond; *c*. side view; *d*. end view; *e*. transverse view.

*** *Frond without a distinct terminal lobe, and frequently having a process or an acute angle at the corners of the terminal portion.*

14. *E. rostratum* (Ralfs); frond oblong; ends protuberant, emarginate and angular, with a horizontal spine on each side.

Euastrum rostratum, Ralfs, *Annals of Nat. Hist.* v. 14. p. 192. t. 7. f. 5. (1844); *Trans. of Bot. Soc. of Edinburgh*, v. 2. p. 129. t. 11. Hass. *Brit. Alg.* p. 383.

Dolgelley and Penzance, *J. R.* Sussex, *Mr. Jenner.* Near Aberdeen, *Dr. Dickie* and *Mr. P. Grant.* Moss Hagg between Tomantaul and Lochavon (alt. 3000 feet), Banffshire, *Mr. P. Grant.* Near Ambleside, *Mr. Sidebotham.* New York, *Bailey.*

Frond very minute, about twice as long as broad; segments obscurely three-lobed, or rather having a broad basal portion, which is emarginate at each side and connects itself by a broad short neck with the terminal lobe. This terminal portion has on each side a horizontal, subacute, beak-like process, and is prominent, emarginate and angular.

Euastrum rostratum, like the species in the preceding section, is contracted (though less decidedly) into a neck, and also has emarginate sides; but it differs from them in its much smaller size, and especially in the horizontal processes of its end lobe. The angular and less prominent ends distinguish it from *E. elegans*.

A single sporangium which I gathered at Dolgelley was spinous, like that of *E. elegans*, but larger.

Length of frond from $\frac{1}{649}$ to $\frac{1}{508}$ of an inch; breadth from $\frac{1}{1000}$ to $\frac{1}{714}$.

Tab. XIV. fig. 6. *a*, *b*. fronds with endochrome.

15. *E. elegans* (Bréb.); frond oblong; ends emarginate, pouting and rounded.

    α. segments slightly constricted beneath the end lobe, which has on each side a short horizontal spine.

    β. *inerme*; segments sinuated rather than lobed, and without spines.

    γ. *spinosum*; segments sinuated rather than lobed, and having two or more spines directed obliquely outwards.

Cosmarium elegans, Bréb. Menegh. *Synop. Desmid. in Linnæa* 1840, p. 222.
Euastrum spinosum, Ralfs, *Annals of Nat. Hist.* v. 14. p. 193. t. 7. f. 6. (1844); *Trans. of Bot. Soc. of Edinburgh*, v. 2. p. 129. t. 11. Hass. *Brit. Algæ*, p. 384.
Euastrum elegans, Kützing, *Phycologia Germanica*, p. 135 (1845).

    Common. Dolgelley, Carnarvon, Penzance, &c., *J. R.* Cheshunt, *Mr. Hassall.* Barmouth, *Rev. T. Salwey.* Near Southampton; Reigate; frequent in Sussex, *Mr. Jenner.* Ireland, *Mr. Andrews.* Hanham near Bristol, *Mr. Thwaites.* Yate near Bristol, *Mr. Broome.* Cheshire and Westmoreland, *Mr. Sidebotham.* Aberdeenshire and Banffshire (alt. 50—3000 feet), *Mr. P. Grant.*

    Falaise, *Brébisson.* Germany, *Kützing.* New York, *Bailey.*

    Frond very minute, and about twice as long as broad; the segments emarginate at their sides, and the ends protuberant, rounded and emarginate.

    The sporangia are orbicular and spinous; they have been gathered at Yate by Mr. Broome, and by myself near Dolgelley.

    Euastrum elegans is a pretty but variable little species; one state of it closely approaches *E. rostratum*, with which it agrees in being somewhat contracted beneath the terminal lobe, and also in its horizontal, though smaller, spines; but all its forms differ from that species by having the prominent ends not angular but rounded.

    The variety β. has no spines and no distinct neck; the sides are undulated rather than lobed, and the ends broadly rounded.

    The variety γ. is distinguished by the spines being directed obliquely forwards, and besides those on the terminal portion it has usually one or two at the sides; there is no distinct neck, and the end is more protuberant and narrow than in the other forms. The spines of its sporangia are more numerous and slender than those in the typal form. I have not seen sporangia from a sufficient number of stations to decide whether differences in their spines indicate different species; but I believe that the number and shape of the spines vary even in the same species.

    Length of frond of α. $\frac{1}{888}$ to $\frac{1}{445}$ of an inch; breadth $\frac{1}{1441}$ to $\frac{1}{714}$; diameter of sporangium $\frac{1}{905}$; length of spine of sporangium $\frac{1}{3086}$. Length of frond of β. $\frac{1}{421}$; breadth $\frac{1}{654}$. Length of frond of γ. $\frac{1}{884}$; breadth $\frac{1}{1388}$; diameter of sporangium $\frac{1}{980}$; length of spine $\frac{1}{2631}$.

    Tab. XIV. fig. 7. *a, b, c.* fronds of α. with endochrome; *d.* sporangium; *e.* frond of var. β; *f.* mature frond of var. γ; *g.* side view; *h.* transverse view; *i, k.* dividing fronds; *l.* sporangium.

16. *E. binale* (Turp.); segments concave or truncate at the end; not projecting beyond the acute angles.

Heterocarpella binalis, Turp. *Dict. des Sc. Nat. par Levr. Alt. Veg.* f. 14. (1820); *Mém. du Mus.* f. 17. Kütz. *Synop. Diat. in Linnæa* 1833, p. 598. Bréb. *Alg. Fal.* p. 56. t. 7.
Cosmarium binale, Menegh. *Synop. Desmid. in Linnæa* 1840, p. 221.
Euastrum binale, Ralfs, *Annals of Nat. Hist.* v. 14. p. 193. t. 7. f. 7. (1844); *Trans. of Bot. Soc. of Edinburgh*, v. 2. p. 130. t. 11. Hass. *Brit. Alg.* p. 384.

Penzance and Dolgelley, *J. R.* Henfield and elsewhere in Sussex; Hampshire, and Reigate, Surrey, *Mr. Jenner*. Hill of Rhoil (alt. 1600 feet), Aberdeenshire, *Dr. Dickie*. Near Aberdeen, *Mr. P. Grant*. Ambleside, *Mr. Sidebotham*. Near Bristol, *Mr. Thwaites*.

Falaise, *Brébisson*. New York, *Professor Bailey*.

Frond very minute, about twice as long as broad; segments inflated at the base, either entire or bicrenate at the sides, slightly contracted upwards and rather dilated at the end. The terminal notch is broad, and forms a concavity between the angles.

The variety β. (which may prove a distinct species) differs in its quadrilateral form and in its truncate ends, which have a small but distinct notch at the centre. The acute angles are slightly prolonged horizontally; the sides of the segments are somewhat crenate, and the frond is rough, with a few scattered and very minute granules.

Euastrum binale is distinguished from the two preceding species by its concave or truncate ends and its notch, the sides of which do not project beyond the lateral spines or angles.

A solitary conjugated specimen was gathered near Bristol by Mr. Broome, but the sporangium was not completely developed.

Length of frond (Tab. XIV. f. 8. *c*.) $\frac{1}{1428}$ of an inch; breadth $\frac{1}{1945}$: length of frond (f. 8. *d*.) $\frac{1}{1179}$; breadth $\frac{1}{1396}$: length of frond (f. 8. *e*.) $\frac{1}{1968}$; breadth $\frac{1}{2403}$: length (f. 8. *f*.) $\frac{1}{1106}$; breadth $\frac{1}{1326}$.

Tab. XIV. fig. 8. *a, c, d, e*. fronds with endochrome; *b*. empty frond; *g*. transverse view; *h*. a specimen conjugating; *f*. frond of var. β.

17. *E. cuneatum* (Jenner); segments cuneate, not lobed; terminal notch not linear.

Euastrum cuneatum, Jenner, *in lit. cum specimine* (1846).

Aberdeenshire, *Dr. Dickie*. Dolgelley, *J. R.* Parham Old Park near Storrington, Sussex, *Mr. Jenner*.

The frond is larger than that of any other species in this section, and about three times longer than broad; the segments are nearly quadrilateral, broadest at the base, and sloping upwards; the ends are truncate, with a short linear notch at the centre. The sides are not sinuated, and I have not detected any inflated protuberances.

The frond has so great a resemblance to some states of *E. crassum*, that, although its segments are not lobed, I think it may prove to be a variety of that species; for in **Dr. Dickie's** specimen I observed a frond, one segment of which answered the above description, whilst the other had a distinct, cuneate, terminal lobe, as in *E. crassum*. But as I lost the frond before I was able to obtain a satisfactory view of it, and as the form here described is so peculiar, and has been noticed in widely separated stations, I have given it the rank of a species, awaiting the result of further observations.

Length of frond $\frac{1}{208}$ of an inch; breadth at base $\frac{1}{420}$; breadth of end $\frac{1}{740}$.
Tab. XXXII. fig. 3. *a.* frond with endochrome; *b.* transverse view.

18. *E.? sublobatum* (Bréb.); segments subquadrate, somewhat contracted beneath the end; the end margin slightly concave.

Euastrum? sublobatum, Brébisson, *in lit. cum icone* (1846).

Machynlleth and Dolgelley, North Wales, *J. R.* Ambleside, *Mr. Sidebotham*.

Falaise, *Brébisson*.

Frond twice as long as broad, and rather larger than that of *E. elegans*; the segments nearly quadrate, their sides and end somewhat sinuated, and their base slightly inflated.

Euastrum sublobatum has nearly as good a claim to a place in *Cosmarium* as in this genus. The terminal notch is here replaced by a slight concavity, and in their form its segments are not unlike those of *Cosmarium quadratum*; but their constricted appearance, from the more strongly-marked lateral concavities, inclines me to retain this species in *Euastrum*.

Length of frond $\frac{1}{523}$ of an inch; breadth at base $\frac{1}{646}$; breadth at constriction $\frac{1}{3246}$; breadth of end $\frac{1}{1028}$.

Tab. XXXII. fig. 4. *a.* frond with endochrome; *b.* empty frond; *c.* transverse view; *d.* end view.

8. COSMARIUM, *Corda*.

Frond simple, constricted in the middle; *segments* as broad as or broader than long, neither sinuated nor notched.

The fronds are minute, simple, constricted in the middle; the segments are generally broader than long and inflato-compressed, but in some species orbicular or cylindrical; they are neither emarginate at the end nor lobed at the sides, and have no spines or processes.

Ehrenberg united plants belonging to this genus with others having lobed segments, in order to form his genus *Euastrum*. Meneghini for the most part followed this arrangement, merely changing the name to *Cosmarium*, which had a prior claim, and also adding some species taken from *Xanthidium*. Under *Micrasterias*, *Euastrum*, and *Xanthidium*, I have given my reasons for differing from such high authorities, and pointed out the characters which distinguish those genera from *Cosmarium*.

Both Ehrenberg and Meneghini consider the inflato-compressed segments essential; but some species with globular or cylindrical segments can by no means be separated from the compressed ones. Those having cylindrical fronds in some respects show an affinity with *Penium*; but in that genus the constriction is either wanting or obscure, and the segments are longer than broad. In *Cosmarium* the fronds are never elongated, are always constricted in the middle, and the starch vesicles are scattered.

Tetmemorus differs from this genus in its elongated fronds and emarginate extremities.

The species belonging to *Cosmarium* may conveniently be divided into three sections.

In the first section the segments are compressed, and in the front view, which differs from the lateral one, they are united by a portion only of their bases, and thus the constriction forms a linear notch on each side. An end view is elliptic.

The plants belonging to the second section are also compressed, and, like those in the preceding, their front view differs from the lateral one and has a linear notch on each side. Each segment however has a central inflation, which causes the end view to be more or less cruciform.

In the third section the end view is circular, the front and lateral views are alike, and there is no linear notch at the sides, since the segments have no constriction at their junction except that which necessarily results from their figure.

* *Frond compressed, deeply constricted at the middle; end view elliptical.*
† *Margin of segments entire.*

1. *C. quadratum* (Ralfs); frond smooth, deeply constricted at the middle; segments in the front view quadrate, and on each side of the base slightly protuberant.

Cosmarium quadratum, Ralfs, *Annals of Nat. Hist.* v. 14. p. 395. t. 11. f. 9. (1844); *Trans. of Bot. Soc. of Edinburgh*, v. 2. p. 151. t. 16. Hass. *Brit. Alg.* p. 367.

Dolgelley, *J. R.* Rusthall Common near Tunbridge Wells, *Mr. Jenner.* Aberdeenshire, *Dr. Dickie* and *Mr. P. Grant.*

Falaise, *Brébisson.*

Frond minute, constricted in the middle, the constriction forming a linear notch on each side. The segments are compressed, and in the front view nearly square; on each side of the base there is a slight protuberance.

C. quadratum is smaller than *C. Cucumis*, and its ends are not so rounded; but as these forms so nearly approach each other, it may prove to be merely a variety. I have however seen no intermediate state, and have therefore kept them distinct.

M. De Brébisson informs me that at the time I described *C. quadratum* in the 'Annals of Natural History,' he had not only recognized it as distinct, but had conferred on it the same specific name.

Length of frond $\frac{1}{510}$ of an inch; breadth $\frac{1}{952}$; breadth at constriction from $\frac{1}{3623}$ to $\frac{1}{3472}$.

Tab. XV. fig. 1. *a.* front view; *b.* side view; *c.* empty frond of variety.

2. **C. Cucumis** (Corda); frond smooth, deeply constricted at the middle; segments as broad as long and rounded at the end; transverse view broadly elliptic.

   β. segments semiorbicular.

Cosmarium Cucumis, Corda, *Alm. de Carlsb.* 1835, p. 121. f. 27. Meneghini, *Synop. Desmid. in Linnæa* 1840, p. 220.
Euastrum integerrimum, Ehr. *Infus.* p. 163. t. 12. f. 9. (1838), according to Meneghini. Kützing, *Phycologia Germanica*, p. 136?
Pithiscus angulosus, Kütz. *Phycol. Germ.* p. 129 (1845).

Midhurst and Fittleworth, Sussex, *Mr. Jenner.* Dolgelley and Penzance, *J. R.* Near Bristol, *Mr. Thwaites.* Aberdeen, *Dr. Dickie.* Banffshire, *Mr. P. Grant.* Ambleside, *Mr. Sidebotham.*

Germany, *Corda, Ehrenberg, Kützing.* Falaise, *Brébisson.* United States, *Bailey.*

Frond minute, about twice as long as broad, compressed, quite smooth, constricted at the middle, the constriction forming a linear notch on each side. In α. the sides are nearly straight, except near the ends, which are rounded; but in β. the sides also are rounded.

C. Cucumis is smaller than *C. Ralfsii*; its frond is not so orbicular, and the transverse view is different.

Length of frond from $\frac{1}{362}$ to $\frac{1}{257}$ of an inch; breadth from $\frac{1}{568}$ to $\frac{1}{502}$; breadth at constriction from $\frac{1}{1436}$ to $\frac{1}{1000}$.

Tab. XV. fig. 2. *a.* front view; *b.* empty frond of var. β; *c.* transverse view.

3. **C. Ralfsii** (Bréb.); frond orbicular, smooth, deeply constricted at the middle; transverse view fusiform.

Cosmarium Cucumis, Ralfs, *in Annals of Nat. Hist.* v. 14. p. 395. t. 11. f. 8. (1844); *Trans. of Bot. Soc. of Edinburgh*, v. 2. p. 151. t. 16. Jenner, *Flora of Tunbridge Wells*, p. 196. Hass. *Brit. Algæ*, p. 366. (Not of Corda according to Brébisson.)
Cosmarium Ralfsii, Brébisson *in lit.* (*cum icone*) (1846).

Dolgelley, *J. R.* Bogs at Fisher's Castle near Tunbridge Wells; Fittleworth near Petworth, Sussex; Hampshire, &c., *Mr. Jenner.* Hill of Rhoil (alt. 1600 feet), Glen Derry, Aberdeenshire, *Dr. Dickie.* Near Aberdeen, *Mr. P. Grant.*

Falaise, *Brébisson.*

Frond large, disciform, quite smooth, deeply constricted at the middle, the constriction forming a linear notch on each side. The segments are semi-orbicular, quite entire, and not punctate.

The endochrome is usually somewhat radiate, and the vesicles are small.

The transverse view is elliptico-lanceolate or fusiform.

This species I formerly considered to be the *Euastrum integerrimum*, Ehr., but M. De Brébisson has convinced me of my mistake. It is a larger plant than that figured by Ehrenberg, and more perfectly orbicular. It differs from *C. Cucumis* by its larger size and disciform appearance.

Length of frond from $\frac{1}{227}$ to $\frac{1}{222}$ of an inch; breadth from $\frac{1}{278}$ to $\frac{1}{261}$; breadth at constriction from $\frac{1}{1250}$ to $\frac{1}{1136}$.

Tab. XV. fig. 3. *a.* frond with endochrome; *b.* transverse view.

4. *C. pyramidatum* (Bréb.); frond oval with flattened ends, deeply constricted in the middle; segments punctate, entire.

Cosmarium ovale, Ralfs, *in Annals of Nat. Hist.* v. 14. p. 394 (in part), t. 11. f. 7. *a, b, c.* (1844); *Trans. of Bot. Soc. of Edinburgh*, v. 2. p. 150 (in part); Jenner, *Flora of Tunbridge Wells*, p. 196.

Cosmarium pyramidatum, Brébisson *in lit. (cum icone)* (1846).

Probably common. Dolgelley and Penzance, *J. R.* Barmouth, *Rev. T. Salwey.* Near Tunbridge Wells; Midhurst; Henfield, &c., Sussex; Reigate, Surrey, *Mr. Jenner.* Near Congleton, Cheshire, and near Ambleside, Westmoreland, *Mr. Sidebotham.* Kerry, *Mr. Andrews.* Hill of Rhoil (alt. 1600 feet), *Dr. Dickie.* Near Aberdeen, *Mr. P. Grant.* Near Manchester, *Mr. Williamson.*

Falaise, *Brébisson.*

Frond about twice as long as broad, varying much in size, subelliptic, with flattened ends, deeply constricted at the middle, the constriction forming a linear notch on each side.

The empty frond is punctate and entire at the margin. Transverse view broadly elliptic.

The sporangia which I have occasionally met with at Dolgelley are orbicular and tuberculated.

From *Cosmarium Cucumis* and *C. Ralfsii* this species may be known by its puncta and depressed ends. From *C. ovale*, with which I formerly united it, it differs in its entire margin and truncated ends. In form the segments approach those of *C. Botrytis*, but are not granulate.

Length of frond from $\frac{1}{471}$ to $\frac{1}{264}$ of an inch; breadth from $\frac{1}{759}$ to $\frac{1}{374}$; breadth at constriction from $\frac{1}{2355}$ to $\frac{1}{1210}$.

Tab. XV. fig. 4. *a, d.* fronds with endochrome; *b.* empty frond; *c, e.* lateral views; *f.* transverse view.

5. *C. tinctum* ——; frond smooth, constriction producing a linear notch on each side; segments elliptic; integument reddish; sporangium naked, subquadrate, conjugating fronds persistent.

Penzance and Dolgelley, *J. R.* Ashdown Forest and Cross-in-hand, Sussex, *Mr. Jenner.*

Frond very minute, smooth, smaller than that of *Cosmarium bioculatum*, deeply constricted at the middle, the constriction causing a linear notch on each side; segments elliptic, twice as broad as long, entire. The empty frond is tinged faint red or straw-colour.

The sporangium is large in proportion to the fronds, and quadrate, with an empty segment of the fronds permanently attached to each corner. Sometimes the fronds couple in a crossed position, when the sporangium appears variously twisted or distorted.

I overlooked this species until I gathered its sporangia near Penzance; and indeed in the growing state it is difficult to distinguish it from *Cosmarium bioculatum*; but it is rather smaller, the notches formed by the constriction are less gaping, and the empty fronds are slightly coloured. The sporangium however differs not only from that of *C. bioculatum*, but from all that I have seen.

Length of frond $\frac{1}{2325}$ of an inch; breadth $\frac{1}{2500}$; length of sporangium $\frac{1}{1602}$.

Tab. XXXII. fig. 7. *a*. front view of frond; *b*. empty frond; *c*. side view; *d*. transverse view; *e, f.* conjugating fronds; *g, h, i.* sporangia.

6. *C. bioculatum* (Bréb.); frond smooth, constriction producing a gaping notch on each side; segments subelliptic, entire; sporangium orbicular, spinous.

Heterocarpella bioculata, Brébisson, *Alg. Fal.* p. 56. t. 7. (1835).
Cosmarium bioculatum, Brébisson *in lit.* (*cum icone*) (1846).

Probably common. Tunbridge Wells, Henfield, &c., Sussex, *Mr. Jenner.* Dolgelley and Penzance, *J. R.* Near Bristol, *Mr. Thwaites.* Near Aberdeen, *Dr. Dickie* and *Mr. P. Grant.* Ambleside, Westmoreland, *Mr. Sidebotham.* Rochdale, *Mr. Coates.*

Falaise, *Brébisson.* United States, *Bailey.*

Frond very minute, about as long as broad, deeply constricted at the middle. The segments, which are elliptic, are connected by a more distinct isthmus than in the allied species, hence there is a wider notch on each side. The endochrome is usually more dense in the centre of each segment. The empty frond is not punctate.

The sporangium is orbicular and minute and has conical spines.

Cosmarium bioculatum differs from *C. Phaseolus*, Bréb. in its smaller size and more elliptic segments, which are not in apposition.

Length of frond $\frac{1}{1416}$ of an inch; breadth of segment $\frac{1}{1773}$; breadth at constriction $\frac{1}{5952}$.

Tab. XV. fig. 5. *a, b.* front view of fronds; *c.* empty frond; *d.* transverse view; *e, f.* sporangia.

7. **C. granatum** (Bréb.); constriction of frond forming a linear notch on each side; segments compressed, smooth, truncato-triangular.

Cosmarium granatum, Brébisson *in lit. cum icone* (1846).

Dolgelley and Penzance, very rare, *J. R.* Bristol, *Mr. Thwaites.*
Falaise, *Brébisson.*

Frond rather larger than those of *Cosmarium bioculatum* and *C. Meneghinii*, twice as long as broad, deeply constricted at the middle, the constriction forming a linear notch on each side; the segments are quite smooth and entire and slope off rapidly from near the base, so that they would be triangular if the end were not truncate.

Not only were the few fronds that I have seen of this species of a larger size than a figure sent me by M. De Brébisson would indicate, but their segments tapered more rapidly. I have seen neither the transverse nor the end view.

This plant may be identical with *Cosmarium Papilio*, Meneghini.

Length of frond $\frac{1}{1234}$ of an inch; breadth of segment at the base $\frac{1}{1602}$; breadth of segment at the end $\frac{1}{5319}$; breadth of frond at constriction $\frac{1}{6111}$.

Tab. XXXII. fig. 6. *a.* mature frond; *b.* empty frond.

†† *Margin of segments crenate; surface not granulate.*

8. **C. Meneghinii** (Bréb.); frond smooth, deeply constricted at the middle; segments subquadrate, sides and end bicrenate.

Cosmarium bioculatum, Menegh. *Consp. Alg. Eug.* p. 18 (1837); *Synop. Desmid. in Linnæa* 1840, p. 220 (not *Heterocarpella bioculata*, Bréb.).
Euastrum bioculatum, Kützing, *Phycologia Germ.* p. 136 (1845).
Cosmarium Meneghinii, Brébisson *in lit. (cum icone)* (1846).

Ambleside, Westmoreland, *Mr. Sidebotham*. Hadlow Down near Mayfield; Henfield; West Chiltington near Pulborough, Sussex; Hants, Kent, and Surrey, *Mr. Jenner*. Dolgelley, *J. R.*

Falaise, *Brébisson*. Italy, *Meneghini*. Germany, *Kützing*. United States, *Bailey*.

The frond is exceedingly minute, quite smooth, and constricted at the middle, the constriction causing a linear notch on each side. The segments are nearly quadrate and the sides and end are bicrenate. The transverse view is elliptic.

Length of frond from $\frac{1}{853}$ to $\frac{1}{735}$ of an inch; breadth from $\frac{1}{1250}$ to $\frac{1}{1176}$; breadth at constriction from $\frac{1}{5555}$ to $\frac{1}{3759}$.

Tab. XV. fig. 6. *a.* front view; *b.* empty frond; *c.* transverse view.

9. **C. crenatum** (Ralfs); frond punctate, deeply constricted at the middle; segments crenate at the margin and flattened at the end; spines of sporangia very short.

Cosmarium crenatum, Ralfs, *in Annals of Nat. Hist.* v. 14. p. 394. t. 11.

f. 6. (1844); *Trans. of Bot. Soc. of Edinburgh*, v. 2. p. 151. t. 16. Jenner, *Flora of Tunbridge Wells*, p. 196. Hassall, *Brit. Alg.* p. 365.

Common. Weston Bogs near Southampton, and several stations in Sussex, *Mr. Jenner.* Dolgelley; Penzance, &c., *J. R.* Hanham near Bristol, with sporangia, *Mr. Thwaites.* Yate near Bristol, *Mr. Broome.* Kerry, *Mr. Andrews.* Near Manchester; and near Ambleside, *Mr. Sidebotham.* Craigendanoch (alt. 1436 feet), Aberdeenshire, *Dr. Dickie.* Hill of Fare (alt. 450 feet), and near Aberdeen, and in Banffshire, *Mr. P. Grant.*

Falaise, *Brébisson.*

Frond smaller than that of *Cosmarium margaritiferum*, twice as long as broad, deeply constricted at the middle, the constriction causing a linear notch on each side. Segments compressed, flattened at the end and crenate at the margin, the surface punctate and even. The end view is elliptic.

The sporangia, which have been gathered abundantly at Bristol by Mr. Thwaites, are orbicular and spinous. The spines, which are very short, and in fact not longer than broad, are divided at the apex and swollen at the base.

The depressed ends distinguish this species from the next.

Length of frond $\frac{1}{474}$ of an inch; breadth at constriction $\frac{1}{1185}$; breadth of segment $\frac{1}{678}$; diameter of sporangium $\frac{1}{856}$; length of its tubercles $\frac{1}{5000}$.

Tab. XV. fig. 7. *a.* mature frond; *b.* empty frond; *c.* sporangium.

10. *C. undulatum* (Corda); frond deeply constricted at the middle; segments semiorbicular, crenate at the margin; sporangia with elongated spines.

Cosmarium undulatum, Corda, *in Almanach de Carlsbad* 1839, p. 243. t. 5. f. 26.

Near Ambleside, *Mr. Sidebotham.* Stapleton near Bristol, *Mr. Thwaites.* Dolgelley, *J. R.* Henfield, &c., Sussex, *Mr. Jenner.*

Carlsbad, *Corda.*

Frond compressed, rather larger than that of *Cosmarium crenatum*, deeply constricted at the middle, the constriction forming a linear notch on each side. The segments are broader than long, crenate or undulate at the margin, the ends rounded. The crenatures are caused by ridges, which are best seen in an end view of the empty frond. The transverse view is elliptic.

Mr. Thwaites has gathered the sporangia near Bristol. They are orbicular, and their spines, which are divided at the apex and swollen at the base, are elongated, and in this respect differ from the last species.

It is not unlikely that some of the habitats given under *Cosmarium crenatum* may belong to this species, since I had confounded these plants until Mr. Thwaites's discovery of the different sporangia.

Length of frond $\frac{1}{416}$ of an inch; breadth of segment $\frac{1}{571}$; diameter of sporangium $\frac{1}{722}$; length of the spines of the sporangium $\frac{1}{3472}$.

Tab. XV. fig. 8. *a, b, d.* front views; *c.* empty frond; *e.* transverse view; *f.* sporangium.

††† *Frond with pearly granules which give a denticulate appearance to the margin.*

11. *C. tetraophthalmum* (Kütz.); frond deeply constricted at the middle; segments semiorbicular, rough with pearly granules which give a crenate appearance to the margin.

Heterocarpella tetrophthalma, Kützing, *Synop. Diatom. in Linnæa* 1833, p. 597. f. 87. Brébisson, *Alg. Fal.* p. 56. t. 7.
Cosmarium ovale, Jenner, *Fl. of Tunbridge Wells,* p. 196 (1845).
Euastrum tetrophthalmum, Kütz. *Phy. Germ.* p. 136 (1845).
Cosmarium tetraophthalmum, Brébisson *in lit.* (1846); not of Meneghini according to Brébisson.

Probably common. Penzance; Dolgelley; and near Carmarthen, *J. R.* Sussex; Hants; and Surrey, *Mr. Jenner.* Near Bristol, *Mr. Thwaites.* Near Ambleside, Westmoreland, *Mr. Sidebotham.* Aberdeen, *Mr. P. Grant.*

Germany, *Kützing.* Falaise, *Brébisson.*

Frond compressed, longer than broad, larger than that of *Cosmarium margaritiferum,* and deeply constricted at the middle, thus acquiring a linear notch on each side. The segments are semiorbicular, or more usually two-thirds of a circle. The pearly granules, by their roundness, produce at the margin a crenate rather than a dentate appearance. The transverse view is broadly elliptic.

Cosmarium tetraophthalmum may be distinguished from *C. Ralfsii* and *C. pyramidatum* by its dentated margin, and from *C. ovale* by its differently-shaped frond and its scattered granules. The pearly granules are, in their form, unlike those of *C. Brebissonii.* It is a larger plant than *C. margaritiferum,* and has also a more rounded margin.

Its sporangia have been gathered by Mr. Thwaites; they are large and their spines finally branched.

Length of frond $\frac{1}{232}$ of an inch; breadth of segment $\frac{1}{326}$; thickness in side view $\frac{1}{479}$; diameter of sporangium $\frac{1}{435}$; length of spines of sporangium $\frac{1}{2732}$.

Tab. XVII. fig. 11. *a.* front view of mature frond; *b.* empty frond; *c.* transverse view. Tab. XXXIII. fig. 8. sporangium with empty fronds attached.

12. *C. ovale* (Ralfs); frond elliptic, deeply constricted at the middle; segments with a marginal band of pearly granules, the disc punctate.

Euastrum, No. 6, Bailey, *Amer. Bacil. in Amer. Journ. of Science and Arts,* v. 41. p. 295. f. 28.
Cosmarium ovale, Ralfs, *in Annals of Nat. Hist.* v. 14. p. 394. t. 11. f. 7. *d, e.* (1844); *Trans. of Bot. Soc. of Edinburgh,* v. 2. p. 15. t. 16. Hassall, *Brit. Algæ,* p. 366.
Euastrum carinatum, Ehr. (Professor Bailey *in lit.* 1846).

Weston Bogs near Southampton, *Mr. Jenner.* Ambleside, Westmoreland, *Mr. Sidebotham.*

West Point, New York, *Bailey.* Falaise, *Brébisson.*

Frond large, oval, twice as long as broad, deeply constricted at the middle; the apposition of the segments for their entire breadth renders the notch on each side linear; the pearly granules are large and confined to the margin, where they form three to six series, the largest and central row producing a dentate appearance; the rest of the frond is merely punctate.

The endochrome of each segment is often longitudinally divided by a pale line. The transverse view is elliptic.

Cosmarium ovale is one of the largest species in the genus; its oval form and its disc devoid of granules sufficiently characterise it.

Length of frond $\frac{1}{139}$ of an inch; breadth of segment $\frac{1}{240}$; breadth of connecting portion $\frac{1}{789}$.

Tab. XV. fig. 9. *a.* front view of empty frond; *b.* side view.

13. *C. Botrytis* (Bory); frond granulate, deeply constricted; segments, in the front view, truncato-triangular; end view elliptic.

Heterocarpella Botrytis, Bory, *Dict. Class.* t. 8 (1825).
Cosmarium deltoides, Corda, *Alm. de Carlsb.* 1835. p. 120. f. 18.
Euastrum Botrytis, Ehr. *Infus.* p. 163 (1838). Kütz. *Phy. Germ.* p. 136.
Euastrum angulosum, Ehr. *Infus.* t. 10. f. 8 (1838).
Cosmarium Botrytis, Meneghini, *Synop. Desm. in Linnæa* 1840, p. 220.
Ralfs, *in Annals of Nat. Hist.* v. 14. p. 393. t. 11. f. 5; *Trans. of Bot. Soc. of Edinburgh*, v. 2. p. 15. t. 16. Hassall, *Brit. Alg.* p. 363.

Dolgelley and Penzance, *J. R.* Bristol, *Mr. Thwaites.* Bexhill, &c., Sussex, *Mr. Jenner.* Manchester; and Ambleside, *Mr. Sidebotham.* Essex, *Mr. Hassall.* Aberdeenshire, *Dr. Dickie* and *Mr. P. Grant.* Banffshire, *Mr. P. Grant.*

Germany, *Ehrenberg.* Falaise, *Brébisson.*

Frond compressed, more or less rough with pearly granules, which give a denticulate appearance to the margin, deeply constricted at the middle, the constriction forming a linear notch on each side. In the front view the segments are broadest at the base and gradually narrower towards the end, which is truncated; hence their figure is somewhat angular. The transverse view is elliptic.

The pearly granules are sometimes very evident, sometimes obscure, but they can always be detected at the margin by the higher powers of the microscope.

Mr. Thwaites has gathered a single sporangium near Bristol; it was orbicular, its spines elongated and slightly divided at the apex.

Cosmarium Botrytis may be known from *C. margaritiferum* by its truncate ends; and from *C. pyramidatum* by its more triangular segments and denticulate margins.

Length of frond from $\frac{1}{469}$ to $\frac{1}{327}$ of an inch; breadth at constriction $\frac{1}{2500}$ to $\frac{1}{1196}$; breadth of segment $\frac{1}{625}$ to $\frac{1}{419}$; diameter of sporangium $\frac{1}{379}$; length of spines of sporangium $\frac{1}{1852}$.

Tab. XVI. fig. 1. *a.* mature frond; *b.* empty frond; *c.* transverse view; *d.* sporangium with empty fronds attached.

14. **C. margaritiferum** (Turp.); frond rough with pearly granules, which are as broad as long; segments semiorbicular or reniform; end view elliptic.

    α. Segments semiorbicular.

    β. *reniformis*; segments reniform.

Ursinella margaritifera, Turp. *Dict. des Sc. Nat.* f. 23 (1820); *Mém. du Mus.* p. 295. t. 13. f. 19.
Cymbella reniformis, Ag. *Consp. Diatom.* p. 10 (1830). Leibl. *Fl.* 1830, p. 315. t. 1. f. 2. Greville, *in Hook. Br. Fl.* v. 2. p. 415. Harvey, *Brit. Alg.* p. 215.
Heterocarpella Ursinella, Kützing, *Synop. Diatom. in Linnæa* 1833, p. 598.
Euastrum margaritiferum, Ehr. *Abh. der Berl. Ak.* 1833, p. 246; *Infus.* p. 163. t. 12. f. 7. Kützing, *Phycol. Germ.* p. 136. Bailey, *Amer. Journ. of Science and Arts*, v. 41. p. 295. f. 8.
Micrasterias margaritifera, Brébisson, *Alg. Fal.* p. 55. t. 7 (1835).
Cosmarium margaritiferum, Meneghini, *Synop. Desmid. in Linnæa* 1840, p. 219. Ralfs, *in Annals of Nat. Hist.* v. 14. p. 393. t. 11. f. 4; *Trans. of Bot. Soc. of Edinburgh*, v. 2. p. 15. t. 16.

Very common. Dolgelley; Carnarvon; near Carmarthen; Penzance, &c., *J. R.* Barmouth, *Rev. T. Salwey.* Sussex; Kent; Surrey; and Weston Bogs near Southampton, *Mr. Jenner.* Farnham, Surrey, *Mr. Reeves.* Herts, *Mr. Hassall.* Near Bristol, *Mr. Thwaites* and *Mr. Broome.* Kerry, *Mr. Andrews.* Near Manchester, *Mr. Gray.* Cheshire and Westmoreland, *Mr. Sidebotham.* Aberdeenshire and Banffshire, *Dr. Dickie* and *Mr. P. Grant.* Rochdale, *Mr. Coates.*

Germany, *Leiblein, Ehrenberg* and *Kützing.* Falaise, *Brébisson.* West Point, New York; and in Mexico, *Bailey.*

Frond compressed, deeply constricted at the middle, the constriction causing a linear notch on each side. The segments are broader than long, semiorbicular or reniform, and rough with pearly granules, which are as broad as long, and give a dentate appearance to the outline. The transverse view is elliptic.

The sporangia of this species, which I have gathered at Dolgelley, are orbicular and inclosed in a granulate cell, which is usually somewhat angular.

Cosmarium margaritiferum varies in size and form. The rough frond distinguishes it from *C. crenatum* and *C. undulatum*; its rounded ends from *C. Botrytis*; its smaller size and shorter granules from *C. Brebissonii*, and its less angular shape from *C. conspersum*.

Length of frond from $\frac{1}{566}$ to $\frac{1}{306}$ of an inch; breadth from $\frac{1}{694}$ to $\frac{1}{416}$; diameter of sporangium $\frac{1}{517}$; length of spines of the sporangium $\frac{1}{2293}$; length of swollen base of spine $\frac{1}{4629}$.

Tab. XVI. fig. 2. *a.* var. β; *b.* empty frond of α; *c.* transverse view; *d.* sporangium with empty segments of conjugated fronds. Tab. XXXIII. fig. 3. *a, b.* mature sporangia.

15. **C. Brebissonii** (Menegh.); frond rough with conic spines or granules; segments semiorbicular; end view elliptic.

Cosmarium Brebissonii, Meneghini, *Synop. Desmid. in Linnæa* 1840, p. 219.

Piltdown Common; Midhurst; Ashdown Forest, near Tunbridge Wells, &c.; Hampshire; Sussex; Reigate, Surrey, *Mr. Jenner.* Dolgelley, *J. R.* Falaise, *Brébisson.*

Frond compressed, larger than that of *Cosmarium margaritiferum*, deeply constricted at the middle, the constriction producing a linear notch on each side. Segments rather broader than long, semiorbicular, rough with pearly granules or spines, which are longer than broad. The transverse view is elliptic.

I am not sure that this form is sufficiently distinct from *Cosmarium margaritiferum*, as the specific difference chiefly depends on the elongated pearly granules, a character which I fear is inconstant.

Our specimens do not agree exactly with a drawing sent me by M. De Brébisson; but in specimens received from him the spines are less crowded than in his figure. He informs me that "Le *Cos. Brebissonii* est plus globuleux que le *C. margaritiferum*, et hérissé de papilles inégales plutôt que de tubercules*."

Length of frond $\frac{1}{285}$ of an inch; breadth of segment $\frac{1}{400}$; length of marginal spine $\frac{1}{7353}$.

Tab. XVI. fig. 3. *a.* empty frond; *b.* side view.

16. *C. conspersum* ———; frond rough with depressed granules; segments quadrilateral; end view elliptic.

Dolgelley, *J. R.* Henfield, *Mr. Jenner.*

Frond compressed, larger than that of *Cosmarium margaritiferum* and having at the middle a deep constriction, which makes a linear notch on each side; segments rather broader than long, quadrilateral, rough with pearly granules, which give a dentate appearance to the outline. The transverse view is elliptic.

Cosmarium conspersum differs from *C. margaritiferum* and *C. Brebissonii* in its quadrilateral figure and in having its granules (which are more depressed than those of *C. Brebissonii*) arranged in lines. In the front view its form is not unlike some states of *C. biretum*; but in that species the granules are smaller, and the end margin is more prominent and generally flattened at the centre.

I formerly considered this plant to be the *C. Brebissonii*, with Meneghini's description of which it certainly agrees better than a specimen so named which has been sent me by M. De Brébisson.

Length of frond $\frac{1}{262}$ of an inch; breadth at constriction $\frac{1}{961}$; breadth of segment $\frac{1}{357}$.

Tab. XVI. fig. 4. *a.* empty frond; *b.* transverse view.

---

* I add Meneghini's characters of the two species:—

"*C. margaritiferum*, cellulis e dorso ellipticis, e latere semiorbicularibus, superficie granulato-margaritifera."

"*C. Brebissonii*, cellulis e dorso elongato-ellipticis, e latere introrsum curvatis, formam trapezii præ se ferentibus, superficie granulato-margaritacea."

17. *C. amœnum* (Bréb.); frond twice as long as broad, with parallel sides and rounded ends, deeply constricted at the middle, rough with pearly granules.

Cosmarium amœnum, Brébisson, *in lit. cum icone* (1846).

Ambleside, Westmoreland, *Mr. Sidebotham.* Dolgelley, *J. R.*
Falaise, *Brébisson.*

Frond smaller than that of *Cosmarium margaritiferum*, twice as long as broad, deeply constricted at the middle, the constriction causing a linear notch on each side; the sides nearly straight and the ends rounded. Segments campanulate, rough with crowded, obtuse, papilla-like, pearly granules which give a dentate appearance to the margin. The side view is narrower and about thrice as long as broad.

In form, *Cosmarium amœnum* resembles the cylindrical species rather than the compressed ones with which it is here classed; but its smaller size and nearly parallel sides will distinguish it from the other species of the latter, whilst from the former it may be known, in the front view, by the linear notch which the constriction produces on each side.

Length of frond $\frac{1}{568}$ of an inch; breadth at constriction $\frac{1}{1470}$; breadth of segment $\frac{1}{1141}$.

Tab. XVII. fig. 3. *a.* mature frond; *b.* empty frond; *c.* side view.

** *Constriction, in the front view, forming a linear notch on each side; end view with a lobe or protuberance on each side.*

† *Frond rough with pearly granules.*

18. *C. biretum* (Bréb.); segments compressed, quadrilateral, broadest at the end; the end margin convex, slightly truncate at the middle; end view with a lobe on each side.

Cosmarium biretum, Brébisson, *in lit. cum icone* (1846).

Hastings and Bexhill, Sussex, *Mr. Jenner.*

Falaise, *Brébisson.*

Frond larger than that of any other species in this section, deeply constricted at the middle, the constriction forming a linear notch on each side. The segments are compressed, quadrilateral or somewhat hexagonal, nearly twice as broad as long, narrowest at the base and gradually dilated upwards. The end margin, which is the longest, is more convex in Sussex than in French specimens, and becomes angular from being slightly flattened at the middle.

The empty frond appears punctate rather than granulate.

An end view is slightly inflated at the middle. The inflation in this and the two following species depends upon a central protuberance of the segment, similar to that observable in some species of *Xanthidium*, and in this respect differs from the inflation of *Cosmarium ornatum* and *C. commissurale*, in which plants it is longitudinal, and extends to or beyond the end margin.

From all except *Cosmarium Broomeii*, this species is known by its quadrangular segments, the lateral margins of which are sloped, not rounded. From that species it differs in its larger size, more distinct angles, and the greater comparative length of the end margin, which is also convex and slightly angular, whilst in *C. Broomeii* it is nearly straight.

Length of frond $\frac{1}{333}$ of an inch; breadth at constriction $\frac{1}{1376}$; breadth of segment $\frac{1}{372}$.

Tab. XVI. fig. 5. *a*. Sussex specimen; *b*. empty frond of French specimen; *c*. end view.

19. **C. Broomeii** (Thwaites); segments compressed, minutely granulate, quadrilateral; end view slightly inflated at the middle.

Cosmarium Broomeii, Thwaites, *in lit. cum specim.* (1846).

In brackish water at Shirehampton near Bristol, *Mr. Broome* and *Mr. Thwaites*. Piltdown, Sussex, *Mr. Jenner*.

Frond as large as that of *Cosmarium margaritiferum*, deeply constricted at the middle, the constriction forming a linear notch on each side. The segments are compressed, minutely granulate, quadrilateral, and twice as broad as long; the end margin is straight.

The end view is twice as long as broad, slightly inflated at the middle and broadly rounded at the ends. In the transverse view the inflation is greater, and there is a slight constriction between it and the ends.

Cosmarium Broomeii resembles in the front view some states of *C. margaritiferum*, but differs in the end one. The inflation does not project at the margin in the front view, as in *C. ornatum*, and the lateral margins are less rounded, the pearly granules are also smaller.

Length of frond $\frac{1}{500}$ of an inch; greatest breadth $\frac{1}{540}$; breadth at constriction $\frac{1}{1562}$; thickness at the inflation in end view $\frac{1}{930}$; least thickness in the end view $\frac{1}{1157}$. Diameter of sporangium $\frac{1}{548}$.

Tab. XVI. fig. 6. *a*. frond with endochrome; *b*. empty frond; *c*. end view; *d*. transverse view. Tab. XXXII. fig. 7. sporangium.

20. **C. cælatum** ——; frond suborbicular; segments crenate, rough with pearly granules; end view slightly inflated at the middle.

Near Machynlleth, North Wales, *J. R.* Tunbridge Wells and Ashdown Forest, Sussex, *Mr. Jenner*.

Falaise, *Brébisson*.

Frond nearly orbicular, scarcely so large as that of *Cosmarium margaritiferum*, compressed, constricted at the middle, the constriction causing a linear notch on each side. The segments are scolloped or broadly crenate: each segment has six crenatures, four at the end, one at each side and one at each intermediate angle, the lateral ones being the largest, whilst between all the crenatures the margin is somewhat sulcated. The pearly granules are distinct, especially on the crenatures, to which they give a dentate appearance. The end view is slightly cruciform, and the transverse view still more so. The lateral view is constricted at the middle and emarginate at the ends.

Cosmarium cœlatum is a very distinct species and requires no comparison with any other.

Length of frond from $\frac{1}{921}$ to $\frac{1}{581}$ of an inch; greatest breadth from $\frac{1}{1024}$ to $\frac{1}{808}$; breadth at constriction from $\frac{1}{2137}$ to $\frac{1}{1538}$.

Tab. XVII. fig. 1. *a.* frond with endochrome; *b.* empty frond; *c.* empty frond of Sussex specimen; *d.* dividing frond; *e.* side view; *f.* end view; *g.* transverse view.

21. *C. ornatum* (Ralfs); segments twice as long as broad, rough with pearly granules which give a dentate appearance to the margin; inflation forming a truncate projection at the end; end view cruciform.

Cosmarium ornatum, Ralfs *in Annals of Nat. Hist.* v. 14. p. 392. t. 11. f. 3 (1844); *Trans. of Bot. Soc. of Edinburgh*, v. 2. p. 148. t. 16. Jenner, *Fl. of Tunbridge Wells*, p. 194. Hassall, *Br. Alg.* p. 364.
Cosmarium Sportella, Brébisson, MS.

Not uncommon. Dolgelley; Barmouth; and Penzance, *J. R.* Ashdown Forest, &c., Sussex, and Weston Bogs near Southampton, *Mr. Jenner.* Cheshunt, *Mr. Hassall.* Kerry, *Mr. Andrews.* Ben Muich Dhu (alt. 2800 ft.); Linn of Dee (1190 ft.); Little Craigendall, Braemar, Aberdeenshire, *Dr. Dickie.* Aberdeen, *Mr. P. Grant.* Ambleside, Westmoreland, *Mr. Sidebotham.*

Falaise, *Brébisson.* West Point, New York, *Bailey.*

The frond is generally smaller than that of *Cosmarium margaritiferum* and deeply constricted at the middle; the constriction forms on each side a linear notch, which is nearly closed by the apposition of the segments. The segments are compressed, about twice as broad as long, and rough with pearly granules which give a dentate appearance to the margin. The inflation extends beyond the margin in a truncate projection, and in the end view forms a rounded lobe on each side, producing a cruciform figure.

Cosmarium ornatum differs from the other species by its inflation extending into a truncate projection beyond the margin.

The sporangium has been found near Bristol by Mr. Broome, and I have gathered it at Penzance. Its spines are elongated, slightly divided at the extremity and dilated at the base.

Length of frond $\frac{1}{613}$ of an inch; greatest breadth $\frac{1}{613}$; breadth at constriction $\frac{1}{2041}$; diameter of sporangium $\frac{1}{658}$; length of spine of sporangium $\frac{1}{1582}$.

Tab. XVI. fig. 7. *a.* front view of mature frond; *b.* empty frond; *c.* end view; *d.* transverse view; *e.* sporangium.

22. *C. commissurale* (Bréb.); segments narrow-reniform, three times broader than long, rough with pearly granules; end view with a constriction between the central inflation and the extremities.

Cosmarium commissurale, Brébisson. Meneghini, *Synop. Desmid. in Linnæa* 1840, p. 220. Brébisson, *in lit. cum icone.*

Piltdown Common, Sussex, *Mr. Jenner.* Near Penzance, *J. R.* Near Bristol, *Mr. Thwaites.*

Falaise, *Brébisson.*

Frond smaller than that of any other species in this section, broader than long, very deeply constricted at the middle. The segments are compressed, three times as broad as long, reniform, sometimes inclining to triangular in consequence of the projection of the inflation at the margin; they are rough with pearly granules which give a dentate appearance to the outline. The transverse view is three times longer than broad, with a constriction between the central inflation and the rounded ends.

The sporangia have been gathered very sparingly by Mr. Jenner. They are orbicular, with slender, elongated spines, which are slightly divided at the apex and dilated at the base.

The great breadth of the frond compared with its length and the reniform segments well distinguish this species.

Length of frond $\frac{1}{923}$ of an inch; greatest breadth $\frac{1}{663}$ to $\frac{1}{609}$; breadth at constriction $\frac{1}{3155}$; thickness of central inflation in end view $\frac{1}{1147}$; diameter of sporangium $\frac{1}{721}$; length of spine of sporangium $\frac{1}{1256}$.

Tab. XVI. fig. 8. *a.* front view of frond; *b.* empty frond; *c.* transverse view; *d.* sporangium.

23. *C. cristatum* ———; frond orbicular, deeply constricted at the middle, margined by papilla-like pearly granules; end view linear, with an inflation at the middle of each side.

Ambleside, Westmoreland, *Mr. Sidebotham.*

Frond exactly orbicular, much compressed, deeply constricted at the middle, the constriction forming a linear notch on each side; the segments margined by a series of obtuse, papilla-like, pearly granules. At the centre of each segment is a circular granulate inflation. The end view is nearly linear, with truncate ends and an inflation on each side at the middle.

The above description is founded on Mr. Jenner's notes and drawings. I have not seen this plant, but have no doubt that it is a good species.

Length of frond $\frac{1}{699}$ of an inch; greatest breadth $\frac{1}{653}$; breadth at constriction $\frac{1}{1785}$.

Tab. XVII. fig. 2. *a.* front view of mature frond; *b.* empty frond; *c.* end view.

†† *Frond smooth.*

24. **C. Phaseolus** (Bréb.); segments smooth, reniform; constriction forming a linear notch on each side; end view elliptic with a slight projection at the middle of each side.

Cosmarium Phaseolus, Brébisson, *in Menegh. Synop. Desmid. in Linnæa* 1840, p. 220. Brébisson, *in lit. cum icone.*

Dolgelley and Penzance, *J. R.*

Falaise, *Brébisson.*

Frond larger than that of *Cosmarium bioculatum,* deeply constricted at the middle, the constriction forming a linear notch on each side; segments twice as broad as long, reniform, quite smooth and entire. The end view is elliptic and shows on each side a slight conical inflation, resembling that seen in *Xanthidium fasciculatum.*

The small circular inflation which is present at the centre of each segment compels me to associate this with the other inflated species; but it differs from all of them in its smooth fronds, and the slight inflation can be detected only by an end view. *Cosmarium Phaseolus* is far more likely to be mistaken for *C. bioculatum,* and indeed is placed next it by Brébisson; but it is larger, its segments are reniform, not elliptic, and the notches caused by the constriction are not gaping as in that species.

Length of frond $\frac{1}{787}$ of an inch; breadth $\frac{1}{833}$; breadth at constriction $\frac{1}{3048}$.

Tab. XXXII. fig. 5. *a.* front view; *b.* empty frond; *c.* end view.

*** *Constriction not forming linear notches at the sides; end view circular.*

† *Frond rough with pearly granules which give a denticulate appearance to the outline.*

25. **C. cylindricum** (Ralfs); segments granulated, subquadrate in the front view, broadest at the extremity; end view circular.

Cosmarium cylindricum, Ralfs, *Annals of Nat. Hist.* v. 14. p. 392. t. 11. f. 1 (1844); *Trans. of Bot. Soc. of Edinburgh,* v. 2. p. 148. t. 16. Jenner, *Fl. of Tunbridge Wells, Supplement,* p. 18. Hass. *Br. Alg.* p. 365.

Wet sides of a cave at Lamorna Cove near Penzance, and wet sides of a cliff at Machynlleth, North Wales, *J. R.* Ashdown Forest; near Tunbridge Wells, and near Battle, *Mr. Jenner.*

Frond minute, cylindrical, about twice as long as broad. The subquadrate segments are broadest at the extremity, which is truncate, somewhat tapering towards the junction, and rough with pearly granules which give a dentate appearance to the outline.

Cosmarium cylindricum may be known from *C. Cucurbita* by the condition of its segments described above.

Length of frond $\frac{1}{588}$ of an inch; greatest breadth $\frac{1}{1060}$; breadth at constriction $\frac{1}{1678}$.

Tab. XVII. fig. 4. *a.* mature frond; *b.* empty frond; *c.* end view.

26. *C. orbiculatum* (Ralfs); segments spherical, rough with pearly granules.

Cosmarium orbiculatum, Ralfs, *Ann. of Nat. Hist.* v. 14. p. 392. t. 11. f. 2 (1844); *Trans. of Bot. Soc. of Edinburgh*, v. 2. p. 148. t. 16. Hass. *Br. Alg.* p. 364.

Dolgelley, *J. R.* Midhurst, Sussex, *Mr. Jenner.* Ambleside, Westmoreland, *Mr. Sidebotham.*

Falaise, *Brébisson.*

Frond minute, composed of two spherical segments, rough with pearly granules, giving a dentate appearance to their outlines, but wanting on the neck-like contraction between them. The transverse view has a large central opening.

Cosmarium orbiculatum varies in size; the larger specimens somewhat resemble the lateral view of *Cosmarium margaritiferum*; the front view however of that species differs in the shape of the segments and in the presence of a linear notch on each side at their junction. The granulated fronds distinguish it from the following species; its central opening also is much larger.

Length of frond from $\frac{1}{498}$ to $\frac{1}{454}$ of an inch; greatest breadth from $\frac{1}{757}$ to $\frac{1}{748}$; breadth at constriction $\frac{1}{1219}$; diameter of sporangium $\frac{1}{833}$; length of spine $\frac{1}{2500}$.

Tab. XVII. fig. 5. *a.* mature frond; *b.* empty frond; *c.* transverse view.
Tab. XXXIII. fig. 9. sporangium.

†† *Frond smooth.*

27. *C. moniliforme* (Turp.); segments spherical, smooth.
   α. Segments united without the intervention of a neck.
   β. Segments united by a distinct neck.

Tessarthronia moniliformis, Turp. *Dict. des Sc. Nat.* t. 7. f. 1. ? (1820).
Tessarthra moniliformis, Ehr. *Abh. d. Berl. Ak.* 1835, p. 173; *Infus.* p. 145. t. 10. f. 20.

α. Dolgelley; rare; *J. R.* Charlton Fields near Manchester, *Mr. Sidebotham.* Ashdown Forest; Fittleworth; and Rackham Bogs near Pulborough, Sussex, *Mr. Jenner.*

β. Penzance.

Germany, *Ehrenberg.*

The frond is small and quite smooth. In α. each segment is a perfect globe, having a very small opening through which it connects with its fellow. Their appearance is one of mere juxtaposition, and their junction can only be ascertained and the opening seen in a transverse view. This state is liable to be mistaken for a not uncommon plant with two globular cells, which I consider to belong properly to the Palmelleæ, and to be the *Scenedesmus moniliformis* of Kützing's 'Phycologia Germanica' and Brébisson's 'Alg. Fal.,' as also the *Trochiscia moniliformis* of Meneghini's 'Synopsis Desmidiearum.' The resemblance indeed is at times so complete that I have occasionally been puzzled to distinguish them; but in specimens of the latter some individuals may generally be perceived which have a slight interval between the globes, whilst its solitary globes are filled with endochrome. In *Cosmarium moniliforme,* as the globes are united they are in close contact, and when separated their contents escape at the orifice.

Both Kützing and Meneghini refer to the *Tessarthra moniliformis* as identical with the plant described by them under the names mentioned above, but I have no hesitation in considering that it is the present plant. Ehrenberg's figure, which exhibits a frond dividing and forming two smaller globes between the original ones, appears to me decisive of the question. M. De Brébisson concurs in this opinion, that the *Tessarthra moniliformis,* Ehr., is not the *Trochiscia moniliformis* of Meneghini's 'Synopsis.'

The var. β. resembles more nearly the form of *Cosmarium orbiculatum,* from having, like that species, a distinct connecting neck; but its fronds are quite smooth and without granules or puncta.

Length of frond $\frac{1}{617}$ of an inch; greatest breadth $\frac{1}{1131}$; breadth at constriction $\frac{1}{2500}$.

Tab. XVII. fig. 6. *a.* usual state; *b.* variety β; *c, d.* dividing fronds; *e.* empty frond; *f.* transverse view.

28. *C. connatum* (Bréb.); segments punctate, each about two-thirds of a circle, uniting by their plane surfaces; end view circular.

Cosmarium connatum, Bréb. *in lit. (cum icone)* (1846).

Ambleside, Westmoreland, *Mr. Sidebotham.* Dolgelley, *J. R.*
Falaise, *Brébisson.*

This is one of the largest species in the genus. The fronds are nearly twice as long as broad, and slightly constricted at the middle. In the front view the segments are broader than long; one-third of the margin is interrupted by their broad junction, and the remainder constitutes two-thirds of a circle; a distinct border is always present, and often appears striated.

The empty frond is punctate rather than granulate, and the end view is circular.

With a little attention this species is easily recognized. Its very broad junction without lateral incisions distinguishes it from those species which have compressed fronds, and its large size and segments broader than long from those which have a circular end view. Its aspect in the front view is more like that of *Staurastrum tumidum,* but the latter has more elliptic segments, and generally shows one at least of its nipple-like angles.

Length of frond $\frac{1}{285}$ of an inch; greatest breadth $\frac{1}{1155}$; breadth at constriction $\frac{1}{1479}$.

Tab. XVII. fig. 10. *a.* mature frond; *b.* empty frond.

29. *C. Cucurbita* (Bréb.); frond punctate, slightly constricted at the middle and rounded at the ends; transverse view circular.

Cosmarium Cucurbita, Bréb. *in Desm. Alg.* No. 1103 (184). Ralfs, *Annals of Nat. Hist.* v. 14. p. 395. t. 11. f. 10; *Trans. of Bot. Soc. of Edinb.*

v. 2. p. 151. t. 16. Jenner, *Fl. of Tunbridge Wells*, p. 196. Hass. *Br. Alg.* p. 367.

Closterium clandestinum, Kütz. *Phy. Germ.* p. 132 (1845). (Kützing *in lit.*)

Rather common. Dolgelley; near Lampeter; and Penzance, *J. R.* Ashdown Forest, Greatham Bogs near Pulborough, Sussex; and near Southampton, *Mr. Jenner*. Near Bristol, *Mr. Thwaites*. Near Congleton, Cheshire; and Ambleside, Westmoreland, *Mr. Sidebotham*. Kerry, *Mr. Andrews*.

Falaise, *Brébisson*. Germany, *Kützing*.

The frond is minute, subcylindrical, twice as long as broad, with a very slight constriction at the middle, and rounded at the ends. The transverse view is circular, with a large opening.

Cosmarium Cucurbita agrees in size with *C. cylindricum*, but is punctate, not granulate. The segments differ also in figure from those of that plant, and resembling two cupping-glasses with their mouths in contact, are aptly characterized by the specific name.

Length of frond $\frac{1}{586}$ of an inch; greatest breadth $\frac{1}{1155}$; breadth at constriction $\frac{1}{1479}$.

Tab. XVII. fig. 7. *a.* front view of mature frond; *b, c.* empty fronds; *d.* transverse view; *e.* dividing frond.

30. **C. Thwaitesii** ———; frond cylindrical, constricted at the middle and rounded at the ends; puncta very indistinct.

Bristol, *Mr. Thwaites*. Swansea, *J. R.*

Frond two to three times longer than broad, larger than that of *Cosmarium Cucurbita*, which in form this species much resembles; sides parallel, the middle slightly constricted and the ends broadly rounded. Mr. Jenner finds very faint, minute, scattered puncta after the escape of the endochrome. Mr. Thwaites's specimens were very gelatinous, but I was unable to ascertain whether each plant had a distinct mucous covering, or was merely imbedded in a common matrix. The transverse view appears to me cylindrical, but Mr. Jenner finds the segments slightly compressed.

Cosmarium Thwaitesii differs from *C. Cucurbita* in its larger size and indistinct puncta; from *C. curtum* in its differently arranged endochrome; and from *C. turgidum* and *C. attenuatum* in its nearly parallel sides.

Length of frond $\frac{1}{357}$ of an inch; greatest breadth $\frac{1}{801}$; breadth at constriction $\frac{1}{833}$.

Tab. XVII. fig. 8. *a.* fronds surrounded with gelatine; *b.* empty frond; *c.* transverse view.

31. **C. curtum** (Bréb.); frond oblong, constricted at the middle and rounded at the ends; endochrome in longitudinal fillets.

Closterium curtum, Brébisson, *in Menegh. Synop. Desmid. in Linnæa* 1840, p. 237. Kützing, *Phycologia Germanica*, p. 132.

Penium curtum, Brébisson, *in lit. cum icone* (1846).

Near Bristol, *Mr. Thwaites*. Penzance, *J. R.*

Falaise, *Brébisson*. Germany, *Kützing*.

Frond minute, smooth, two to three times longer than broad, slightly constricted at the middle, attenuated and rounded at the ends. Endochrome in three or four longitudinal fillets, which are interrupted at the suture. End view circular, its endochrome with four geminate rays.

This plant was placed in *Penium* by M. De Brébisson because its endochrome is arranged in fillets, but in every other respect it agrees better with some of the preceding species of this genus. It is more obtuse than the usual form of *Cosmarium attenuatum*, but I have occasionally seen specimens of the latter so nearly resembling the figure of *C. curtum* that I have doubted to which species they belonged. It is much smaller than *C. turgidum*, and tapers more than *C. Cucurbita* and *C. Thwaitesii*.

Length of frond $\frac{1}{465}$ of an inch; greatest breadth $\frac{1}{1064}$.

Tab. XXXII. fig. 9. *a.* mature frond; *b.* empty frond; *c.* end view of frond with-stellate endochrome.

32. *C. attenuatum* (Bréb.); frond fusiform, three or four times longer than broad, slightly constricted at the middle; transverse view circular.

Cosmarium attenuatum, Brébisson, *in lit. cum icone* (1846).

Near Bristol, *Mr. Thwaites.* Swansea, *J. R.*

Falaise, *Brébisson.*

Frond three or four times longer than broad, tapering, its ends obtuse; end view circular. The empty frond is finely punctate.

Cosmarium attenuatum is much smaller and more tapering than *C. turgidum*, and has less-rounded ends. Its fusiform frond distinguishes it from *C. Thwaitesii* and *C. Cucurbita*. In general it tapers more than *C. curtum*, but sometimes its ends are more obtuse, and then these species can only be distinguished by the different arrangement of their endochromes.

It is doubtful whether this plant and *Cosmarium turgidum* ought not to be removed to *Penium*.

Length of frond from $\frac{1}{420}$ to $\frac{1}{416}$ of an inch; greatest breadth from $\frac{1}{1068}$ to $\frac{1}{1022}$; breadth at constriction from $\frac{1}{1142}$ to $\frac{1}{1099}$.

Tab. XVII. *a, b.* fronds in usual state; *c.* empty frond; *d.* transverse view.

33. *C. turgidum* (Bréb.); frond oblong, punctate, constricted at the middle and rounded at the ends; transverse view circular.

Cosmarium turgidum, Brébisson, *in lit. cum icone* (1846).

Henfield, *Mr. Jenner.* Swansea, *J. R.*

Falaise, *Brébisson.*

Frond large, oblong, turgid, about three times as long as broad, distinctly constricted at the middle and broadly rounded at the ends. Empty frond with numerous, minute, scattered puncta.

Cosmarium turgidum differs from the allied species in its much greater size. It is comparatively longer and more tapering than *C. Cucurbita* and *C. Thwaitesii*, and its ends are more rounded than those of *C. attenuatum*.

Length of frond $\frac{1}{126}$ of an inch; greatest breadth $\frac{1}{249}$; breadth at constriction $\frac{1}{271}$.

Tab. XXXII. fig. 8. *a.* mature frond; *b.* empty frond; *c.* transverse view.

9. XANTHIDIUM, *Ehr.*

Frond simple, constricted in the middle; *segments* compressed, entire, spinous, having a circular projection near the centre, which is usually tuberculated.

In this genus the frond is simple and deeply constricted at the middle; its segments are slightly compressed, turgid, entire, either reniform, orbicular or angular; they are furnished with spines, which, simple or branched, are either scattered or confined to the margin; in the latter case they are arranged in two rows, one on each side of the marginal line. Near the centre of each segment is a remarkable circular projection on both surfaces. These projections are usually surrounded by a beaded circle of pearly granules or tubercles, which are most evident in a lateral or transverse view.

Xanthidium requires to be distinguished from three genera, viz. *Staurastrum, Arthrodesmus,* and *Cosmarium*. Its resemblance to *Staurastrum* is more apparent than real, and will probably mislead no one who has examined a plant belonging to that genus, in which the cells are angular irrespective of the processes. In *Arthrodesmus* each segment has only two spines, one on each side, and there are no central projections.

The connection with *Cosmarium* is far more intimate; in fact the sole distinctive character that can be relied upon is the presence of spines in this genus. Meneghini indeed still further reduces the number and value of the differential marks, by retaining in *Xanthidium* merely those species whose spines are scattered over the surface, whilst he refers to *Cosmarium* those in which the spines are confined to the margin. I think however that his views cannot be admitted, and that we must either retain the genus as formed by Ehrenberg, or unite all the species with *Cosmarium*; for in *X. armatum* the principal spines are marginal, and many specimens, especially in a young state, have none other.

I have followed preceding writers in taking the specific characters chiefly from the form and position of the spines; but Mr. Jenner considers that the central protuberances will afford better distinctive marks of some species. However, I have seen too few examples to decide this question.

One plant (*X. octocorne,* Ehr.) without central projections, and in other respects different from the true species of *Xanthidium,* I have retained here, because I am unable to assign it a more proper situation.

* *Spines divided at the apex.*

1. *X. armatum* (Bréb.); segments broadest at the base; spines short, stout, terminated by three or more diverging points.

Cosmarium armatum, Brébisson, *Menegh. Synop. Desmid. in Linnæa* 1840, p. 218.
Xanthidium furcatum, Ralfs, *in Annals of Nat. Hist.* v. 14. p. 466. t. 12. f. 1 (1844), excluding synon.; *Trans. of Bot. Soc. of Edinburgh,* v. 2. p. 154. t. 17. Jenner, *Fl. of Tunbridge Wells,* p. 194. Hassall, *Brit. Alg.* p. 359.
Euastrum armatum, Kützing, *Phy. Germ.* p. 137 (1845).
Xanthidium armatum, Brébisson, *in lit.* (1846).

Common. North Wales; near Carmarthen; and Penzance, *J. R.* Sussex; Kent; Weston Bogs near Southampton, and New Forest, *Mr. Jenner.* Herts, *Mr. Hassall.* Kerry, *Mr. Andrews.* Ireland, *Mr. Moore.* Aberdeenshire, *Mr. P. Grant* and *Dr. Dickie.* Banffshire, *Mr. P. Grant.* Ambleside, Westmoreland, *Mr. Sidebotham.*

Falaise, *Brébisson.* Germany, *Kützing.*

This species forms, at the bottom of shallow pools, cloud-like masses, which being detached, immediately rise to the surface. Its fronds, dispersed like minute glittering dots through the connecting cloud-like substance, are visible to the naked eye.

Frond comparatively large, deeply constricted, the constriction forming a linear notch on each side; segments nearly equal in length and breadth, broadest at the base, their ends rounded or truncate; spines short, stout, divided at the apex; of these there are generally six pairs, which are marginal, and situated three on each side; a few others usually are scattered on the disc.

The central protuberances are cylindrical, truncate, and bordered by pearly granules which produce a dentate appearance, especially in the end and lateral views. The empty frond is minutely punctate. The end view is elliptic.

I have gathered a single sporangium of *Xanthidium armatum* at Dolgelley; it was large and orbicular, with depressed tubercles. Having seen but this example, I am unable to say whether these tubercles eventually elongate into spines.

Xanthidium Artiscon, Ehr., differs from this species not only in its smaller size and its differently-shaped segments, but also in its longer spines, which are nearer the ends than those of *X. armatum.*

Length of frond $\frac{1}{180}$ of an inch; breadth $\frac{1}{270}$; thickness $\frac{1}{280}$; breadth at constriction $\frac{1}{625}$.

Tab. XVIII. *a.* frond with endochrome; *b.* side view; *c, d.* fronds acquiring new segments by division; *e.* empty frond; *f.* transverse view; *g.* sporangium with empty segments of the conjugated fronds.

** *Spines subulate.*

2. **X. aculeatum** (Ehr.); spines subulate, more or less scattered; central projections truncate, obscurely dentate.

Xanthidium aculeatum, Ehr. *Abh. d. Berl. Ak.* (1833), p. 318; *Infus.* p. 147. t. 10. f. 23. Meneghini, *Synop. Desmid. in Linnæa* 1840, p. 224. Ralfs, in *Annals of Nat. Hist.* v. 14. p. 467. t. 12. f. 2; *Trans. of Bot. Soc. of Edinburgh*, v. 2. p. 155. t. 17. Jenner, *Fl. of Tunbridge Wells*, p. 194. Hassall, *Brit. Alg.* p. 360.

Weston Bogs near Southampton; Ashdown Forest, and Tunbridge Wells, *Mr. Jenner*. Penzance, *J. R.* Yate near Bristol, *Mr. Thwaites*.

Germany, *Ehrenberg*.

Frond deeply constricted at the middle, so as to form a linear notch on each side; segments somewhat reniform, twice as broad as long; spines subulate, short, marginal and scattered. The central projections, as seen in end or lateral views, are cylindrical, truncate, and bordered by pearly granules which give a dentate appearance, but less conspicuously than in *Xanthidium armatum*.

Xanthidium aculeatum varies in the arrangement and number of its spines. The Bristol specimens have a far larger central projection than the other specimens which I have seen.

This species is distinguished from all the following by having its spines more or less scattered.

Length of frond, not including spines, from $\frac{1}{384}$ to $\frac{1}{377}$ of an inch; including spines, from $\frac{1}{324}$ to $\frac{1}{302}$; breadth, not including spines, from $\frac{1}{393}$ to $\frac{1}{347}$; including spines, from $\frac{1}{316}$ to $\frac{1}{270}$; breadth at constriction from $\frac{1}{1529}$ to $\frac{1}{1165}$.

Tab. XIX. fig. 1. *a, e, g.* fronds with endochrome; *b,* empty frond; *c, h.* side views; *d, f.* end views.

3. **X. Brebissonii** ——; spines subulate, marginal, geminate; central projection somewhat truncate and margined with pearly granules.

β. *varians*; segments broader and more irregular; spines somewhat irregular and unequal.

Binatella aculeata, Brébisson, *Alg. Fal.* p. 58. t. 8 (1835).
Cosmarium aculeatum, Meneghini, *Synop. Desmid. in Linnæa* 1840, p. 218 (in part).
Euastrum, No. 11, Bailey, *Amer. Bacil. in Amer. Journ. of Science and Arts*, v. 41. t. 1. f. 13 (1841).
Xanthidium bisenarium, Ehrenberg, *Verb. und Einfluss des mikrosk. Lebens in Süd- und Nord-Amer.* (1843).
Xanthidium aculeatum, Brébisson, *in lit. cum icone* (1846).

α. Piltdown Common, *Mr. Jenner*. Bristol, *Mr. Thwaites*. Penzance, *J. R.*
β. Piltdown Common, Sussex, *Mr. Jenner*. Trewellard near Penzance, *J. R.*
Falaise, *Brébisson*. West Point, New York, *Bailey*.

Frond deeply constricted at the middle; segments broader than long; spines marginal. In α. all the spines are geminate and equal, but in β. they are more or less unequal and not so regularly geminate, and occasionally also a few are scattered. Mr. Jenner finds, both in the Sussex and the Cornish specimens of this variety, a single spine interposed between the two central pairs and the lateral ones. This I have also observed, but not invariably. The central projections resemble those of *Xanthidium aculeatum*, but the pearly granules are smaller. I believe that every species of *Xanthidium* except *X.? octocorne* is punctated, but the puncta can be detected only by the highest powers of the microscope.

By Brébisson this species is considered the *X. aculeatum*, Ehr., but Ehrenberg's description and figure represent the spines as scattered. According to Professor Bailey's figure in his 'American Bacillaria,' this plant is probably the *Xanthidium bisenarium*, Ehr., but that name is inapplicable to our forms. The number of spines seems to be variable. Bailey's and Ehrenberg's figures have six pairs of spines, Brébisson's eight, whilst in the Sussex specimens there are generally ten, to each segment.

Length of frond, not including spines, $\frac{1}{416}$ of an inch; including spines, $\frac{1}{320}$; breadth, not including spines, $\frac{1}{408}$; including spines, $\frac{1}{314}$; breadth at constriction $\frac{1}{1360}$.

Length of var. β, not including spines, $\frac{1}{413}$; including spines, $\frac{1}{333}$; breadth, not including spines, $\frac{1}{365}$; including spines, $\frac{1}{268}$; breadth at constriction $\frac{1}{1298}$.

Tab. XIX. fig. 2. *a.* front view of typal form; *b.* dividing frond of var. β; *c.* empty frond; *d.* end view.

4. *X. fasciculatum* (Ehr.); segments with 4–6 pairs of subulate, marginal spines; central projections minute, conical, not beaded.

α. Each segment with four pairs of spines.

Xanthidium fasciculatum, Ehr. *Infus.* p. 146. t. 10. f. 24. *a.* (1838). Ralfs, in *Annals of Nat. Hist.* v. 14. p. 467. t. 12. f. 3. *a, b, c, d*; *Trans. of Bot. Soc. of Edinburgh*, v. 2. p. 154. t. 17. Jenner, *Fl. of Tunbridge Wells*, p. 194.
Cosmarium antilopæum, Brébisson, in Menegh. *Synop. Desmid. in Linnæa* 1840, p. 218.
Euastrum, No. 10, Bailey, *Amer. Bacil. in Amer. Journ. of Science and Arts*, v. 41. p. 296. t. 1. f. 10 (1841).
Euastrum fasciculatum, Kützing, *Phy. Germ.* p. 137 (1845).
Xanthidium polygonum, Hassall, *Brit. Freshwater Alg.* p. 360, excluding synonyms (1845).
Xanthidium fasciculatum, var. antilopæum, Brébisson, *in lit. cum icone* (1846).

β. *polygonum*; six pairs of spines to each segment.

Xanthidium fasciculatum β. *polygonum*, Ehr. *Infus.* t. 10. f. 24. *b.* (1838). Ralfs, in *Annals of Nat. Hist.* v. 14. t. 12. f. 3. *e*; *Trans. of Bot. Soc. of Edinburgh*, v. 2. p. 154.
Xanthidium fasciculatum, Hassall, *Brit. Alg.* p. 359 (1845).

α. Common. Dolgelley and Penzance, *J. R.* Sussex; and near Southampton, *Mr. Jenner.* Yate near Bristol, *Mr. Broome.* Kerry, *Mr. Andrews.*

Aberdeenshire, *Dr. Dickie* and *Mr. P. Grant*. Cheshire; and Ambleside, Westmoreland, *Mr. Sidebotham*. Rochdale, *Mr. Coates*.

β. Penzance and Dolgelley, *J. R.*

Germany, *Ehrenberg* and *Kützing*. Falaise, *Brébisson*. West Point, New York, *Bailey*.

Frond smaller than that of *Xanthidium aculeatum*, deeply constricted at the middle; segments twice as broad as long; spines subulate, slender, geminate and marginal, in general slightly curved, but occasionally straight. Each segment commonly has only four pairs, though sometimes six. The central projections are conical, inconspicuous, to be detected only in end and lateral views, and even then liable to be overlooked. A mucous covering can usually be detected in α, but I am doubtful of its presence in β.

The variety β. Mr. Jenner considers nearer to *Xanthidium Brebissonii* than to *X. fasciculatum*; but, in my opinion, it agrees better with this species in size, and especially in the shape of the central projections.

Length of frond, not including spines, from $\frac{1}{454}$ to $\frac{1}{350}$ of an inch; including spines, from $\frac{1}{341}$ to $\frac{1}{274}$; breadth, not including spines, from $\frac{1}{432}$ to $\frac{1}{408}$; including spines, from $\frac{1}{290}$ to $\frac{1}{282}$; breadth at constriction $\frac{1}{1533}$.

Length of var. β, not including spines, $\frac{1}{481}$; including spines, $\frac{1}{384}$; breadth, not including spines, $\frac{1}{516}$; including spines, $\frac{1}{370}$; breadth at constriction $\frac{1}{1923}$.

Tab. XX. fig. 1. *a.* front view of frond with endochrome; *b.* front view of frond with mucous covering; *c.* empty frond; *d.* transverse view; *e.* lateral view.

Tab. XIX. fig. 4. var. β. *a.* front view of frond with endochrome; *b.* empty frond; *c.* lateral view; *d.* end view.

5. *X. cristatum* (Bréb.); segments with a solitary spine on each side at the base, the other spines geminate.

a. Segments reniform; spines scarcely curved.

Xanthidium cristatum, Brébisson, *in lit.* (1846).

β. *uncinatum*, Bréb.; segments truncate at the end; spines uncinate.

Xanthidium uncinatum, Brébisson, *in lit. cum icone* (1846).
Xanthidium cristatum β. *uncinatum*, Brébisson, *in lit.* (1846).

α. Not uncommon. Dolgelley and Penzance, *J. R.* Ambleside, *Mr. Sidebotham*. Aberdeen, *Mr. P. Grant*. Pulborough, Sussex, *Mr. Jenner*.

β. Ambleside, Westmoreland, *Mr. Sidebotham*. Dolgelley, *J. R.*

α. and β. Falaise, *Brébisson*.

Frond smaller than that of *Xanthidium fasciculatum*, the spines subulate and marginal; each segment has four pairs, and on each side of the base a solitary one; the central elevations, which are less distinctly granulate in α. than in β, are scarcely visible in a front view, but in an end one form slight, lateral, conical projections, the pearly granules encircling them like a necklace. In α. the segments are subreniform and all their spines nearly straight. The segments in β. are about equal in length and breadth, the ends being truncated, hence they are somewhat trapezoid; their basal spines are nearly straight, the rest are curved.

This species is sufficiently marked by its solitary basal spines.

Length of frond, not including spines, $\frac{1}{357}$ of an inch; including spines, $\frac{1}{353}$; breadth, not including spines, $\frac{1}{499}$; including spines, $\frac{1}{353}$; breadth at constriction $\frac{1}{1666}$.

Length of frond of var. *uncinatum*, not including spines, $\frac{1}{469}$; including spines, $\frac{1}{344}$; greatest breadth, not including spines, $\frac{1}{625}$; including spines, $\frac{1}{454}$; breadth at end $\frac{1}{1470}$; breadth at constriction $\frac{1}{2289}$.

Tab. XIX. fig. 3. *a*. front view of typal form; *b*. empty frond; *c*. end view; *d*. front view of var. *uncinatum*; *e*. empty frond; *f*. end view.

6. *X.? octocorne* (Ehr.); segments much compressed, without a central protuberance, trapezoid, each angle terminated by one or two spines.

    a. One spine at each angle.

    β. *major*; larger; two or more spines at each angle.

Arthrodesmus octocornis, Ehr. *Infus*. p. 152 (1838). Hassall, *Brit. Alg.* p. 357. Brébisson, *in lit.*

Micrasterias octocornis, Meneghini, *Synop. Desmid. in Linnæa* 1840, p. 216.

Cosmarium palmatum, Brébisson, *according to Menegh. in Synop. Desmid.* (1840).

Staurastrum? octocorne, Ralfs, *in Annals of Nat. Hist.* v. 15. p. 159. t. 12. f. 3 (1845); *Trans. of Bot. Soc. of Edinburgh*, v. 2. p. 145. t. 15.

Xanthidium octocorne, Ehr. *Meteorp.* t. 1. f. 22.

Euastrum octocorne, Kützing, *Phy. Germ.* p. 134 (1845).

    α. Dolgelley and Penzance, *J. R.* Yate near Bristol, *Mr. Broome*. Ambleside, Westmoreland, *Mr. Sidebotham*. Pulborough, Sussex; Reigate, Surrey; and Weston Bogs near Southampton, *Mr. Jenner*.

    β. Dolgelley, *J. R.*

    α. Germany, *Ehrenberg* and *Kützing*. Falaise, *Brébisson*. Worden's Pond, Rhode Island, *Bailey*.

Frond minute, deeply constricted at the middle; segments trapezoid, each angle terminated by one or more straight spines, the intervals between the angles concave.

The variety β, which I have gathered mixed with the more common form, is usually larger, and its appearance, on account of the geminate spines, differs so much from α. that Mr. Jenner regards it as another species; but I believe that no dependence can be placed on the number of spines as a specific distinction, for in specimens which I sent to Mr. Jenner, he discovered one frond whose lateral spines were single and end ones geminate, and other fronds with three spines on each angle of one segment and two on each angle of the other; some again with the lateral spines geminate and the end spines in threes. I have, in one instance, seen the spines double on one segment and single on the other.

The proper position of this plant is so doubtful, that different observers have referred it to widely different genera. Ehrenberg placed it first in *Arthrodesmus* and afterwards in *Xanthidium*; Brébisson considered it a *Cos-*

marium and more recently an *Arthrodesmus*; whilst Meneghini and Kützing unite it with the *Micrasteriæ*. It differs from the true species of *Xanthidium*, not only by usually having its spines in a single series, but also by its more compressed segments, and by the absence of a central projection; for which reasons Mr. Jenner proposes to make it the type of a new genus, connecting as it were *Xanthidium* with *Arthrodesmus* or *Staurastrum*. Since there is so much diversity of opinion, I have thought it advisable to keep it in *Xanthidium* until its proper situation can be determined with greater certainty.

Length of frond $\frac{1}{1351}$ of an inch; greatest breadth $\frac{1}{1538}$; breadth at constriction $\frac{1}{5000}$.

Length of frond of var. β. $\frac{1}{1020}$ of an inch; greatest breadth $\frac{1}{906}$; breadth at constriction $\frac{1}{2331}$.

Tab. XX. fig. 2. *a*. front view of frond with endochrome; *b*. empty frond; *c*. end view; *d, e*. fronds acquiring new segments by division; *f*. front view of a frond of var. β. with endochrome; *g*. empty frond; *h*. side view; *i*. end view.

10. ARTHRODESMUS, *Ehr.*

Frond simple, compressed, constricted at the middle; *segments* entire, with a single spine on each side.

The fronds are simple, much compressed and deeply constricted at the middle; the segments, which are broader than long, have a single spine or mucro on each side, but are otherwise smooth and entire.

The sporangia of one species only have hitherto been observed; they are spinous.

Where the plants should be placed, to the reception of which I have restricted this genus, has been left in much uncertainty. Ehrenberg, making no distinction between constricted and binate cells, has associated them with others belonging to *Scenedesmus* to form his *Arthrodesmus*; Meneghini assigns them to *Staurastrum*, and Kützing to *Euastrum*. In my papers in the 'Annals of Natural History,' I followed the arrangement of Meneghini, but at the same time suggested that these plants should be united with *Arthrodesmus octocornis*, Ehr., in a distinct genus, for which Ehrenberg's name might be retained. This suggestion Mr. Hassall has adopted in his 'British Freshwater Algæ'; and it is approved by Brébisson. Mr. Jenner however considers that they belong to *Staurastrum*; and if he is correct in his supposition that he has met with one species some specimens of which were compressed, whilst others had three angles in an end view, the former arrangement ought to be restored. It must further be allowed, that in the front view the resemblance to some species of *Staurastrum* is very close.

1. *A. convergens* (Ehr.); segments elliptic, each having its spines curved towards those of the other.

Arthrodesmus convergens, Ehr. *Infus.* p. 152. t. 10. f. 18 (1838). Hassall, *Brit. Algæ*, p. 357.

Staurastrum convergens, Meneghini, *Synop. Desmid. in Linnæa* 1840, p. 228. Ralfs, *in Annals of Nat. Hist.* v. 15. p. 158. t. 12. f. 1; *Trans. of Bot. Soc. of Edinburgh*, v. 2. p. 145. t. 15. Brébisson, *in lit.*

Euastrum, No. 12, Bailey, *in Amer. Journ. of Science and Arts*, v. 41. p. 296. t. 1. f. 11 (1841).

Euastrum convergens, Kützing, *Phy. Germ.* p. 136 (1845).

Dolgelley and Penzance, *J. R.* Sussex and Kent, *Mr. Jenner.* Herts, *Mr. Hassall.* Yate near Bristol, *Mr. Broome.* Aberdeenshire, *Dr. Dickie* and *Mr. P. Grant.* Kerry, *Mr. Andrews.* Near Congleton, Cheshire; and Ambleside, Westmoreland, *Mr. Sidebotham.* Rochdale, *Mr. Coates.*

Germany, *Ehrenberg, Kützing.* Falaise, *Brébisson.* West Point, New York, *Bailey.*

Frond smooth and deeply constricted at the middle; the transversely elliptic segments have on each side a curved spine which converges with the similar one of the other segment.

The frond has a gelatinous covering, which is sometimes distinctly seen, sometimes imperceptible.

Arthrodesmus convergens differs from *A. Incus* by its larger size and elliptic segments.

Length of frond from $\frac{1}{598}$ to $\frac{1}{539}$ of an inch; breadth, not including spines, from $\frac{1}{584}$ to $\frac{1}{477}$; including spines, from $\frac{1}{394}$ to $\frac{1}{384}$; breadth at constriction $\frac{1}{2368}$.

Tab. XX. fig. 3. *a.* frond with endochrome; *b.* empty frond; *c.* end view; *d.* frond dividing.

2. *A. Incus* (Brèb.); segments with end margin truncate.

    a. Segments externally lunate; spines diverging.

    β. Segments gibbous on each side near the base; spines of the one segment parallel to or converging with those of the other.

Cosmarium Incus, Brébisson (1839), *according to Meneghini in Synop. Desmid.*

Staurastrum Incus, Meneghini, *Synop. Desmid. in Linnæa* 1840, p. 228. Ralfs, *in Annals of Nat. Hist.* v. 15. p. 158. t. 12. f. 2; *Trans. of Bot. Soc. of Edinburgh*, v. 2. p. 145. t. 15.

Euastrum, Bailey, *in Amer. Journ. of Science and Arts*, v. 41. t. 1. f. 12? (1841).

Euastrum retusum, Kützing, *Phy. Germ.* p. 136 (1845), (according to Kützing *in lit. cum icone*).

Arthrodesmus Incus, Hassall, *Brit. Freshwater Algæ*, p. 357 (1845).

Dolgelley and Penzance, *J. R.* Weston Bogs near Southampton; and several stations in Sussex, *Mr. Jenner.* Yate near Bristol, *Mr. Broome.* Aberdeenshire, *Mr. P. Grant* and *Dr. Dickie.* Near Ambleside, Westmoreland, *Mr. Sidebotham.*

Falaise, *Brébisson.* Germany, *Kützing.* West Point, New York, *Bailey.*

Frond smooth, minute, deeply constricted at the middle. The fronds are truncated at the ends and have a mucro at each angle; a distinct neck at the constriction is sometimes present, sometimes wanting. In α. the segments are semilunate and the spines usually directed outwards. The variety β. is generally larger than α; its segments are nearly equal in length and breadth, and being rather suddenly rounded off at the constriction, present a gibbous-like appearance on each side; the spines in β. are stouter, and those of one segment slightly converge with those of the other. This form is considered a distinct species by Mr. Jenner, who thinks that its end view is sometimes triangular, and that it unites *Arthrodesmus* with *Staurastrum*.

The orbicular sporangia have been gathered in Sussex by Mr. Jenner, and I have met with them at Dolgelley and Penzance; they have subulate spines like those of *Staurastrum dejectum*, which they resemble, except in their smaller size.

This species is variable both in the form of its segments and in the direction of its spines; the ends however are always truncate; a character which affords a good distinction between it and *A. convergens*.

Euastrum Incus of Kützing's 'Phycologia Germanica' (*Arthrodesmus minutus*, Bréb. MS.) appears, from a drawing which he sent me, to differ merely in the smaller size and in its parallel spines. I fear it is scarcely distinct.

Arthrodesmus Incus, in the front view, has much the appearance of some states of *Staurastrum dejectum*; it is nevertheless usually smaller, and its ends are more decidedly truncate.

Length of frond (Tab. XX. fig. 4. *a*.) $\frac{1}{1103}$ of an inch; breadth $\frac{1}{1020}$; length of frond (fig. 4. *e*.) from $\frac{1}{1666}$ to $\frac{1}{1361}$; breadth, not including spines, from $\frac{1}{1960}$ to $\frac{1}{1420}$; including spines, from $\frac{1}{946}$ to $\frac{1}{793}$; breadth at constriction from $\frac{1}{3846}$ to $\frac{1}{3521}$; diameter of sporangium, without spines, $\frac{1}{1210}$; including spines, $\frac{1}{753}$.

Length of frond of var. β. $\frac{1}{833}$; breadth, not including spines, $\frac{1}{1116}$; including spines, $\frac{1}{537}$; breadth at constriction $\frac{1}{2777}$.

Tab. XX. fig. 4. *a, e.* fronds with endochrome; *b.* empty frond; *c, h.* end views; *d.* frond dividing; *i.* fronds conjugating; *k, l.* sporangia; *f.* frond of var. β. with endochrome; *g.* empty frond.

11. STAURASTRUM, *Meyen.*

Frond simple, constricted at the middle; *end view* angular, or circular with a lobato-radiate margin, or, rarely, compressed with a process at each extremity.

Frond mostly minute, simple, more or less constricted at the middle, and thus forming two segments, which are often somewhat twisted, generally broader than long, and in many species elongated laterally into a process, so that the constriction on each side is a roundish or angular sinus; in other respects the front view shows the segments quite entire.

The end view varies in form: in most of the species it is triangular or quadrangular, and the angles are either rounded or elongated into rays; in some it is circular with five or more processes forming marginal rays; in a few species it is compressed and the extremities terminate by a process.

Ehrenberg in his great work has distributed the plants, which I shall here describe, among different genera according to the number of angles or processes seen in an end view. Thus he refers those with three angles to *Desmidium,* and those with four to *Staurastrum,* and he formed his genus *Pentasterias* for the reception of a plant with five rays. But this arrangement appears unnatural; not only because it separates nearly allied forms, but also because the number of rays, as Meneghini remarks, is not constant even in the same species: Professor Bailey says, when describing an American species, "The number of arms is usually three, but I have met with specimens in which one corpuscle had three and the other four arms, others in which both had four, and others again in which both had five arms." I have myself seen fronds of *Staurastrum paradoxum* and *S. dejectum,* one segment of which had four, and the other only three rays. I have generally found the *Pentasterias margaritaceum* of Ehrenberg with six rays, although not unfrequently with five, and occasionally with seven rays to each segment. Since we already know that about half the British species vary in the number of their rays, and specimens with three rays conjugate with four-rayed ones, the number of rays is not only altogether useless as a generic character, but it does not distinguish species and scarcely indeed varieties.

Staurastrum contains more species than any other genus in the family. M. de Brébisson enumerates forty in a list which he has sent me, to which several must be added that have been figured by Professor Bailey and others. The species exhibit a great variety of forms, and but little affinity can be traced between many of them:— as, for instance, between *Staurastrum tumidum* and *S. paradoxum,* or between *S. dejectum* and *S. bacillare.* Nevertheless, desirable as it would be to separate plants so little allied, it seems better to keep them together for the present than prematurely to attempt their separation.

I should greatly have preferred Kützing's name *Phycastrum* to *Staurastrum*; for the latter is applicable only to a limited number of the species. Still, as it had been previously employed by Meneghini and Brébisson, the rule established in such cases scarcely permits its exchange for the more appropriate term.

Sporangia of several species have been gathered. They are generally spinous, but differ in other respects.

A little care will distinguish *Staurastrum* from the other genera in this family, although some of its species appear at first sight to approach forms which belong to them. It differs from *Desmidium* in never forming a filament; and from *Arthrodesmus* and *Cosmarium* by its angular shape, or by having the ends elongated into processes. Some species bear a considerable resemblance to species of *Xanthidium*, to which genus Ehrenberg refers them; but in *Xanthidium* the frond, irrespective of the spines, is not angular in the end view, and there is a projection at the centre of each segment in the front view.

* *Frond smooth, or rough with minute puncta-like granules; end view with the lobes or angles inflated and mucronate or awned.*

† *Frond smooth.*

1. *S. dejectum* (Bréb.); segments smooth, lunate or elliptic; constricted portion very short; end view with inflated awned lobes.

a. Segments externally lunate, awns directed outwards.

Staurastrum dejectum, Bréb., Meneghini, *Synop. Desmid. in Linnæa* 1840, p. 227. Brébisson, *in lit. cum icone.*

Staurastrum mucronatum β, Ralfs, *in Annals of Nat. Hist.* v. 15. p. 152 (1845). t. 10. f. 5.

β. Segments elliptic, awns parallel.

γ. Awns converging.

Staurastrum mucronatum α. and γ, Ralfs, *in Annals of Nat. Hist.* v. 15. p. 152. t. 10. f. 5; *Trans. of Bot. Soc. of Edinburgh,* v. 2. p. 139. t. 13. Jenner, *Fl. of Tunbridge Wells,* p. 192.

Goniocystis (*Trigonocystis*) *mucronata,* Hass. *Brit. Alg.* p. 350 (1845).

Common. Dolgelley and Penzance, *J. R.* Weston Bogs near Southampton; Sussex; Surrey; and Kent, *Mr. Jenner.* Kerry, *Mr. Andrews.* Yate near Bristol, *Mr. Broome.* Aberdeenshire, *Dr. Dickie* and *Mr. P. Grant.* Ambleside, *Mr. Sidebotham.* Rochdale, *Mr. Coates.*

Falaise, *Brébisson.* West Point, New York, *Bailey.*

Frond smooth, deeply constricted at the middle; segments broader than long, lunate or elliptic, awned, the awns subulate, varying in length and direction. The end view shows three or four mammillate lobes or rays, each terminated by a hair-like mucro or awn.

Sporangia of this plant are more commonly met with than those of any other species in the family except *Hyalotheca dissiliens.* Since I first detected them at Dolgelley, I have gathered them more or less abundantly every year, both at Dolgelley and at Penzance; they have also been sent me by Mr. Jenner, Mr. Broome, Dr. Dickie and Mr. Thwaites.

The conjugated fronds are connected by the formation of a bag-like receptacle or cell, which is colourless and very thin, and therefore difficult of detection. As this enlarges the fronds become more remote from each other, their segments partially separate at the constriction on the inner side, and the endochromes of both pass out and unite to form an orbicular body between them. In this state it resembles the sporangium formed by some species of *Closterium*. At first it is inclosed in an orbicular membrane much larger than itself; but as it increases in density, fine hairs make their appearance on the surface and gradually become stout spines; the membrane then disappears, and the sporangium acquires its perfect state covered with conspicuous awl-shaped spines. At this stage the empty fronds seem scarcely connected with the sporangium, except that they lie on opposite sides of it, have their openings towards it, accompany its movements, and always retain the same relative position.

The above description of the process applies, with occasional and slight variations, to the conjugation of the greater number of species in the Desmidieæ whose sporangia have been noticed; and I have detailed it at length in this place, not only because this species has afforded more numerous opportunities for tracing the formation of its sporangium, but because it was the first example I had witnessed of the spinous kind, and I devoted more time to its examination than to any other.

Either more than one species has been included under this name, or the sporangia vary much in regard to their spines, which in the Penzance specimens are fewer and appear at first like minute tubercles.

Staurastrum dejectum and a few other species form a distinct group, distinguished by their smooth fronds, the peculiar inflated or mammillate form of their angles, or rather lobes, in the end view, and by their terminal awn-like spines. In some respects they have more resemblance to the two plants placed in *Arthrodesmus* than to the other species of this genus. Should any change be required, I would rather remove them to *Arthrodesmus* than unite that genus with *Staurastrum*.

Staurastrum dejectum is larger than *S. cuspidatum*, its spines are shorter, and its segments are connected either without a band or by a very short one.

Length of frond $\frac{1}{833}$ of an inch; breadth $\frac{1}{757}$; breadth at constriction $\frac{1}{2732}$; length of awn $\frac{1}{3937}$; diameter of sporangium $\frac{1}{714}$; length of spine of sporangium from $\frac{1}{2500}$ to $\frac{1}{1250}$.

Tab. XX. fig. 5. *a, b, c.* front views of frond; *d, e.* end views; *f, g.* conjugated fronds; *h, i, k, l, m.* different states of sporangia.

2. *S. cuspidatum* (Bréb.); segments smooth, fusiform, connected by a long narrow band; awns parallel or converging, but straight; end view with inflated awned lobes.

Binatella tricuspidata, Brébisson, *Alg. Fal.* p. 57. t. 8 (1835).
Staurastrum cuspidatum, Bréb., Meneghini, *Synop. Desmid. in Linnæa* 1840, p. 226. Brébisson, *in lit. cum icone*.
Phycastrum cuspidatum, Kützing, *Phy. Germ.* p. 138 (1845).

Dolgelley and Penzance, *J. R.* Rochdale, *Mr. Coates.* Weston Bogs near Southampton, *Mr. Jenner.*

Falaise, *Brébisson.* Germany, *Kützing.*

Frond usually smaller than that of *Staurastrum dejectum*; segments smooth, fusiform or truncate on the outer side, separated by a long band-like neck which is present also in the dividing frond between the old and new segments; the subulate awns are elongated and straight, and either parallel to those of the other segment or convergent with them. The end view has three or four mammillate and awned lobes.

Staurastrum cuspidatum may be known from *S. dejectum* by the elongated band which connects the segments.

The sporangium of this species resembles that of *S. dejectum*, but has fewer spines.

Length of frond $\frac{1}{883}$ of an inch; breadth $\frac{1}{1000}$; breadth at constriction $\frac{1}{3937}$; diameter of sporangium $\frac{1}{869}$.

Tab. XXI. fig. 1. *a, b.* fronds with endochrome; *c.* frond dividing; *d.* transverse view; *e.* end view.

Tab. XXXIII. fig. 10. sporangium.

3. **S. aristiferum** ——; segments smooth, the lobes in front view prolonged into mammillate awned projections which are somewhat constricted at the base; end view with three or four awned lobes.

Dolgelley, *J. R.*

Frond smooth, about as large as that of *Staurastrum cuspidatum*; segments not separated by a band; ends usually truncate, but the angles are elongated into mammillate processes, each terminated by a long awn; end view as in the two preceding species.

Staurastrum aristiferum may be a variety of *S. cuspidatum,* but as it has no isthmus between the segments, and the angles in the front view are elongated, I have described it as distinct.

Length of frond $\frac{1}{657}$ of an inch; breadth $\frac{1}{1064}$; length of spine $\frac{1}{1923}$; breadth at constriction $\frac{1}{2500}$.

Tab. XXI. fig. 2. *a.* front view; *b.* end view.

4. **S. Dickiei** ——; segments smooth, subelliptic, turgid; spines short, curved toward those of the other segment; end view with three slightly inflated mucronate lobes.

Aberdeenshire, *Dr. Dickie.* Dolgelley, *J. R.*

Frond smooth, larger than that of *Staurastrum dejectum,* deeply constricted at the middle; segments twice as broad as long, elliptic, but having the outer margin more convex than the inner one; spines short, curved and converging with those of the other segment. End view with three slightly inflated lobes, which are rounded and mucronate.

This plant was first sent me by Dr. Dickie, who pointed out its resemblance in a front view to *Arthrodesmus convergens.* It is however smaller, although it has an additional angle, and I have not detected any mucous covering. From the three preceding species it differs in its more turgid segments and in the short, curved, converging spines. As the sides are but slightly concave, the end view is less decidedly lobed, the angles also are more rounded and less conspicuously mammillate. *Staurastrum Dickiei* is more like the *S. brevispina*, Bréb., but that species has larger segments, which are semiorbicular and very turgid, and its spines are very minute, and detected, at least in the dried specimen, with difficulty. The Scotch specimens have their spines smaller than those of the Welsh ones, and therefore approach still more closely to *S. brevispina*.

Length of frond $\frac{1}{855}$ of an inch; breadth $\frac{1}{929}$; length of awn $\frac{1}{3086}$; breadth at constriction $\frac{1}{2500}$.

Tab. XXI. fig. 3. *a.* front view; *b.* end view.

5. **S. brevispina** (Bréb.); segments smooth, turgid, elliptic, minutely mucronate; end view three-lobed, each lobe terminated by a short mucro.

Staurastrum brevispina, Brébisson, Meneghini, *Synop. Desmid. in Linnæa* 1840, p. 229. Brébisson, *in lit. cum icone et specimine*.

Henfield, and near Pulborough, Sussex, *Mr. Jenner*. Penzance, *J. R.* Falaise, *Brébisson*.

Frond larger than that of *Staurastrum dejectum*; segments elliptic or reniform, very turgid, twice as broad as long, having on each side a very minute spine or mucro which is often difficult to detect. A drawing sent me by M. de Brébisson represents the spines subulate and converging with those of the other segment. In Mr. Jenner's specimens there is a minute papilla rather than mucro at each lobe, and in the front view these papillæ are situated more outwardly than the spines are in M. de Brébisson's specimens. The end view is three-lobed; the lobes are inflated and have broadly rounded ends.

Length of frond $\frac{1}{502}$ of an inch; breadth $\frac{1}{510}$; breadth at constriction $\frac{1}{2041}$; length of spine $\frac{1}{12195}$.

Tab. XXXIV. fig. 7. *a.* front view; *b.* end view; *c, d.* front and end views from drawings by M. de Brébisson.

†† *Frond rough with minute granules.*

6. **S. lunatum** ——; frond rough with puncta-like granules; segments externally lunate, with an awn at each angle; end view with three inflated awned lobes.

Penzance, *J. R.*

This species resembles in figure *Staurastrum dejectum*, but it is larger. Frond deeply constricted at the middle; segments semilunate, the convex margins united, the outer margin rough with minute granules and truncate, each angle tipped by an awn or mucro which is directed obliquely outwards. End view three-lobed, the lobes inflated, obtuse and awned.

Its rough frond distinguishes *Staurastrum lunatum* from all the preceding species, and the inflated awned lobes of its end view from all the following ones.

Length of frond $\frac{1}{850}$ of an inch; breadth $\frac{1}{680}$; breadth at constriction $\frac{1}{2336}$; length of spine $\frac{1}{4098}$.

Tab. XXXIV. fig. 12. *a.* front view; *b.* end view.

** *Frond smooth; angles in end view broadly rounded.*

7. **S. *muticum*** (Bréb.); segments smooth, elliptic, end view showing slightly concave sides and 3—5 rounded angles.

Binatella muticum, Brébisson, *Alg. Fal.* p. 57. t. 8 (1835).
Staurastrum trilobum, Meneghini, *Conspect. Alg. Eug.* p. 18 (1837).
Staurastrum muticum, Bréb. *in Menegh. Synop. Desmid. in Linnæa* 1840; *in lit. cum icone.*

Dolgelley and Penzance, *J. R.* Ambleside, *Mr. Sidebotham.*
Falaise, *Brébisson.* Italy, *Meneghini.*

The frond, generally smaller than that of *Staurastrum orbiculare*, is quite smooth and has a mucous covering, which is sometimes distinct, but often it can only be ascertained by the appearance of a clear border round the frond which prevents the contact of other substances; segments elliptic. The end view shows three or four angles, sometimes five, according to Meneghini; the sides are slightly concave and the angles broadly rounded.

Staurastrum muticum differs from *S. orbiculare* by its elliptic segments and mucous covering.

Length of frond $\frac{1}{674}$ of an inch; breadth $\frac{1}{680}$; breadth at constriction $\frac{1}{2500}$.

Tab. XXI. fig. 4. *a, b.* front view; *c.* end view of three-rayed state; *d.* end view of four-rayed variety.

Tab. XXXIV. fig. 13. Sporangium, from a drawing by M. de Brébisson.

8. **S. *orbiculare*** (Ehr.); segments smooth, semiorbicular; end view bluntly triangular.

Desmidium orbiculare, Ehr. *Abh. der Berl. Ak.* 1832, p. 292; *Infus.* p. 141. t. 10. f. 9.
Staurastrum orbiculare, Ralfs, *in Annals of Nat. Hist.* v. 15. p. 152. t. 10. f. 4. (1845); *Trans. of Bot. Soc. of Edinburgh*, v. 2. p. 138. t. 13. Jenner, *Fl. of Tunbridge Wells*, p. 96. Brébisson, *in lit. cum icone.*
Phycastrum orbiculare, Kützing, *Phy. Germ.* p. 137 (1845).
Goniocystis (Trigonocystis) orbicularis, Hassall, *Brit. Freshwater Algæ*, p. 349 (1845).

Dolgelley and Penzance, *J. R.* Sussex; Kent; and Weston Bogs near Southampton, *Mr. Jenner.* Ben Muich Dhu (alt. 3480 feet), Aberdeenshire, *Dr. Dickie.* Aberdeenshire and Banffshire, *Mr. P. Grant.* Ambleside, Westmoreland, *Mr. Sidebotham.*

Germany, *Ehrenberg* and *Kützing.* Falaise, *Brébisson.* West Point, New York, *Bailey.*

Frond smooth, deeply constricted at the middle; the segments are generally in close approximation for their entire breadth, hence the frond is suborbicular. The end view is triangular, with sides either straight or slightly concave and rounded angles.

Staurastrum orbiculare may be recognized by its smooth frond, angles broadly rounded and destitute of processes, and its orbicular form in a front view.

Sporangia, gathered at Dolgelley and Penzance, are orbicular, with subulate spines. The conjugating fronds are smaller than those in the ordinary state.

Length of frond $\frac{1}{1037}$ of an inch; breadth $\frac{1}{1106}$; breadth at constriction $\frac{1}{3205}$; diameter of sporangium $\frac{1}{960}$; length of spine $\frac{1}{1730}$.

Tab. XXI. fig. 5. *a, c, d, h.* front view; *b, f, i.* end view; *e.* transverse view; *g.* sporangium.

9. **S. tumidum** (Bréb.); segments smooth, elliptic or suborbicular; end view bluntly triangular, each angle terminated by a nipple-like projection.

Binatella tumida, Brébisson, *Alg. Fal.* p. 66 (1835).
Staurastrum orbiculare, Meneghini, *Synop. Desmid. in Linnæa* 1840, p. 225.
Staurastrum tumidum, Brébisson, *in lit. cum icone* (1846).

Piltdown Common, Sussex, *Mr. Jenner*. Dolgelley and Penzance, *J. R.*

Falaise, *Brébisson*.

Frond large and distinctly visible to the naked eye, smooth, deeply constricted at the middle, having a distinct gelatinous covering, which is often marked with close radiating striæ; segments elliptic or suborbicular. The end view is triangular or quadrangular, the sides are convex, the angles rounded, and each terminated by a nipple-like process; some of these projections are visible in the front view. In both views the margin of the frond appears striated. The empty frond is minutely punctate.

Length of frond $\frac{1}{200}$ of an inch; breadth $\frac{1}{250}$; breadth at constriction $\frac{1}{450}$.

Tab. XXI. fig. 6. *a.* front view; *b.* end view of triangular form; *c.* end view of quadrangular variety magnified only 200 times; *d.* transverse view.

*** *Frond with simple spines, hairs, or (rarely) acute granules; angles, in end view, broadly rounded and entire.*

10. **S. muricatum** (Bréb.); segments semiorbicular, rough with conic granules; end view triangular, with convex sides and broadly rounded angles.

Binatella muricata, Brébisson, *Alg. Fal.* p. 66 (1835).
Desmidium apiculosum, Ehrenberg, *Infus.* p. 142 (1838). Pritchard, *Infus.* p. 184.
Staurastrum muricatum, Brébisson, Meneghini, *Synop. Desmid. in Linnæa* 1840, p. 226.
Xanthidium deltoideum, Corda, *Observ. Microscopiques sur les Animalcules de Carlsbad*, p. 29. t. 5. f. 38, 39 (1840).
Staurastrum muricatum β, Ralfs, *Annals of Nat. Hist.* v. 15. p. 154. t. 11. f. 1. *d, e.* (1845); *Trans. of Bot. Soc. of Edinburgh*, v. 2. p. 141. t. 14.
Phycastrum apiculosum, Kützing, *Phy. Germ.* p. 137 (1845).

Goniocystis (Trigonocystis) muricata β, Hassall, *Brit. Freshwater Algæ*, p. 351 (1845).

Sussex, *Mr. Jenner.* Dolgelley, *J. R.*
Falaise, *Brébisson.* Germany, *Ehrenberg, Corda, Kützing.*

Frond deeply constricted at the middle, nearly equal in length and breadth, rough with minute conic granules or spines. End view triangular, its sides convex and angles broadly rounded.

Staurastrum muricatum is larger than *S. hirsutum*, and not hirsute, but rough with stout short granules or spines; in the end view also the sides are more convex.

Length of frond $\frac{1}{400}$ of an inch; breadth $\frac{1}{474}$.

Tab. XXII. fig. 2. *a.* front view; *b.* end view.

11. **S. hirsutum** (Ehr.); segments semiorbicular, rough with numerous hair-like spines; end view with three rounded angles and straight or slightly convex sides.

Xanthidium hirsutum, Ehrenberg, *Abh. der Berl. Ak.* 1833, p. 318; *Infus.* p. 147. t. 10. f. 22. Meneghini, *Synop. Desmid. in Linnæa* 1840, p. 224.
Binatella hispida, Brébisson, *Alg. Fal.* p. 58. t. 8 (1835).
Xanthidium pilosum, Ehr. *Bericht der Berl. Ak.* 1836 (according to Meneghini).
Staurastrum muricatum, Ralfs, *Annals of Nat. Hist.* v. 15. p. 154. t. 11. f. 1. *a, b, c* (1845); *Trans. of Bot. Soc. of Edinburgh*, v. 2. p. 141. t. 14. Jenner, *Fl. of Tunbridge Wells*, p. 194.
Euastrum (Xanthidium) hirsutum, Kützing, *Phycologia Germanica*, p. 137 (1845).
Goniocystis (Trigonocystis) muricata, Hassall, *Brit. Freshwater Algæ*, p. 351 (1845).
Staurastrum hirsutum, Brébisson, *in lit. cum icone* (1846).

Dolgelley and Penzance, *J. R.* Sussex, *Mr. Jenner.* Ambleside, *Mr. Sidebotham.* Rochdale, *Mr. Coates.* Yate, *Mr. Broome.* Aberdeenshire, *Mr. P. Grant.*

Germany, *Ehrenberg, Kützing.* Falaise, *Brébisson.*

Frond variable in size, about equal in length and breadth, deeply constricted at the middle; segments semiorbicular, hirsute rather than spinous; hairs numerous, scattered. End view triangular; the sides straight or slightly convex, and the angles broadly rounded.

I have gathered sporangia at Dolgelley and Penzance. They are orbicular, their spines short, and branched at the apex.

The hair-like spines are characteristic of this species.

Length of frond from $\frac{1}{676}$ to $\frac{1}{468}$ of an inch; breadth from $\frac{1}{833}$ to $\frac{1}{680}$; breadth at constriction $\frac{1}{2300}$; diameter of sporangium from $\frac{1}{744}$ to $\frac{1}{480}$; length of spine of sporangium $\frac{1}{2040}$.

Tab. XXII. fig. 3. *a.* front view of frond with endochrome; *b.* front view of empty frond; *c.* end view of frond with endochrome; *d.* end view of empty frond; *e.* transverse view; *f.* immature sporangium; *g, h.* mature sporangia.

12. *S. teliferum* ——— ; segments reniform, bristly ; end view triangular, with concave sides and broadly-rounded bristly angles.

Dolgelley and Penzance, *J. R.* Ambleside, *Mr. Sidebotham.* Hadlow Down near Mayfield, and several other stations in Sussex, *Mr. Jenner.*

Frond about as large as that of *Staurastrum hirsutum*, deeply constricted at the middle ; segments twice as broad as long, somewhat reniform, and furnished with scattered spines. End view triangular ; the spines variable in number and confined to the angles.

Staurastrum teliferum differs from *S. hirsutum* in its longer spines, which are also fewer, stouter, and in the end view confined to the angles. It is a larger plant than *S. Hystrix* ; its spines are more numerous, and the end margins in the front view are convex.

Length of frond $\frac{1}{597}$ of an inch ; breadth $\frac{1}{643}$; breadth at constriction $\frac{1}{2041}$; length of spine $\frac{1}{4098}$; diameter of sporangium $\frac{1}{738}$; length of spine of sporangium $\frac{1}{2066}$.

Tab. XXII. fig. 4. *a.* front view ; *b.* end view.
Tab. XXXIV. fig. 14. Sporangium.

13. *S. Hystrix* ——— ; segments subquadrate, spinous ; end view with three or four rounded angles, each furnished with a few subulate spines.

Dolgelley, *J. R.* Near Storrington, Sussex, *Mr. Jenner.*

Frond very minute, smaller than that of any other species with simple spines, deeply constricted at the middle ; segments spinous, somewhat quadrilateral, about twice as broad as long, the end margin nearly straight. The spines are subulate, few in number, and in the front view most of them are lateral. The end view is triangular or quadrangular, and has concave sides and rounded angles. Six or eight spines are scattered on each angle, the sides and disk being naked.

Its smaller size and differently shaped segments will distinguish this species from *Staurastrum teliferum*.

Length of frond from $\frac{1}{1075}$ to $\frac{1}{1020}$ of an inch ; breadth from $\frac{1}{1165}$ to $\frac{1}{954}$; breadth at constriction $\frac{1}{1165}$; length of spine $\frac{1}{6024}$.

Tab. XXII. fig. 5. *a.* front view of frond with endochrome ; *b.* front view of empty frond ; *c.* end view ; *d.* transverse view of four-rayed variety.

**** *End view of frond showing four or more toothed lobes, which are either truncate or rounded, but never elongated into rays.*

14. *S. quadrangulare* (Bréb.) ; frond smooth ; segments quadrangular, with a few marginal spines or teeth ; end view quadrilateral, with truncate angles either emarginate or dentate.

Staurastrum quadrangulare, Brébisson, *in lit. cum icone* (1846).

Ambleside, *Mr. Sidebotham.*
Falaise, *Brébisson.*

Frond very minute, smooth, deeply constricted at the middle, the constriction forming a linear notch on each side; segments quadrilateral, twice as broad as long, with one or two minute but stout teeth at each angle. End view quadrangular, the sides concave, the angles truncate, and in British specimens emarginate.

A drawing of *Staurastrum quadrangulare* sent me by Brébisson represents a much larger form. Its angles, which in the end view are broader, have four teeth at their truncate ends and two minute teeth on their upper surface; but I concur with Mr. Jenner in regarding the British plant as a small variety of this species.

Length of frond $\frac{1}{1157}$ of an inch; breadth $\frac{1}{1163}$; breadth at constriction $\frac{1}{2193}$.

Tab. XXII. fig. 7. *a*. front view with endochrome; *b*. empty frond; *c*. end view.

Tab. XXXIV. fig. 11. *a*. front view from a drawing by M. de Brébisson; *b*. end view.

15. *S. sexcostatum* (Bréb.); segments in the front view with a toothed angle at each side; end view circular, with five or six broad, short, toothed lobes.

Staurastrum sexcostatum, Brébisson, Meneghini, *Synopsis Desmid. in Linnæa* 1840, p. 228; Brébisson, *in lit. cum icone*.
Staurastrum Jenneri, Ralfs, *Annals of Natural History*, v. 15. p. 158. t. 11. f. 8 (1845); *Transactions of Bot. Society of Edinburgh*, v. 2. p. 144. t. 14. Jenner, *Fl. of Tunbridge Wells*, p. 194.
Goniocystis (*Pentasterias*) *Jenneri*, Hassall, *Brit. Freshwater Algæ*, p. 356 (1845).

Very rare. Between Mayfield and Hadlow Down, Sussex, *Mr. Jenner*. Yate near Bristol, *Mr. Thwaites*. Dolgelley, *J. R.*

Falaise, *Brébisson*.

Frond large, rough with conic granules which give a dentate appearance to the outline; segments about as broad as long, produced into a toothed angle on each side, where also a triangular sinus is formed between the angles. The end view is circular and elevated in the centre, and has five or six broad, short, toothed, marginal lobes. The transverse view has a large central opening surrounded by a row of large granules.

I refer this plant to Brébisson's *Staurastrum sexcostatum*, on his own authority, but a drawing sent by him of *S. sexcostatum* represents a smaller state.

Length of frond $\frac{1}{661}$ of an inch; breadth from $\frac{1}{833}$ to $\frac{1}{694}$; breadth at constriction $\frac{1}{1597}$.

Tab. XXIII. fig. 5. *a*. front view of Sussex specimen; *b*, *c*. end views; *d*. transverse view; *e*. front view of Dolgelley specimen.

***** *Frond smooth; end view acutely triangular, with two accessory subulate spines to each angle.*

16. *S. monticulosum* (Bréb.); segments with a forked spine on each side, and at the end about four short, stout, acute projections; end view acutely triangular, with a bifid appendage to each angle.

Staurastrum monticulosum, Brébisson, Menegh. *Synop. Desmid. in Linnæa* 1840, p. 226.
Phycastrum monticulosum, Kützing, *Phycologia Germanica*, p. 138 (1845).

Penzance; very rare, *J. R.*

Falaise, *Brébisson.*

Frond smooth, rather large, deeply constricted at the middle; segments somewhat elliptic, with a projection on each side forked like the tail of a swallow. The end margin has generally four remarkable projections, which are stout, twice as long as broad, acute, and look not unlike a cluster of pyramids. The end view is triangular, the angles acute. At the base of each angle, on the upper surface, is a bifid projection, the acute points of which appear on each side.

Staurastrum monticulosum is a remarkable species, difficult to describe, but easily recognized by its figure*.

Tab. XXXIV. fig. 9. *a.* front view of French specimen; *b.* end view; both from drawings by M. de Brébisson.

17. *S. pungens* (Bréb.); frond smooth, each end with about six subulate spines, directed outwards; each angle in the end view tapering into a spine, which has two smaller ones at its base.

Euastrum, No. 14, Bailey, *Amer. Journ. of Science and Arts*, v. 41. p. 297. t. 1. f. 14? (1841).
Staurastrum pungens, Brébisson, *in lit. cum icone* (1846).

Penzance; rare, *J. R.* Cross-in-Hand, Sussex, *Mr. Jenner.*

Falaise, *Brébisson.* West Point, New York, *Bailey.*

Frond smooth, deeply constricted at the middle; segments externally lunate, and having on the outer margin about six subulate spines, the lateral ones being the most conspicuous. The end view has three angles which taper gradually into subulate spines; its upper surface has two smaller spines at the base of each angle, which diverge and become visible on opposite sides.

Staurastrum pungens agrees with *S. monticulosum* in having two spines on the upper surface of each angle; but its spines are more slender, and in the front view are all simple and directed outwards.

Tab. XXXIV. fig. 10. *a.* front view; *b, c.* end views.

---

* " *S. monticulosum*, Bréb., cellulis e dorso triangularibus, apicibus acutissimis, suprapositione bifidis, prominentia bifido-divaricata breviore, quoque angulo superimposita, prominentiis in centro confluentibus, a latere ellipticis utrinque bifidis, exterius dentatis."—*Meneghini.*

****** *Frond smooth; front view with diverging processes divided at the apex.*

18. **S. brachiatum** ——; frond smooth; front view with thick diverging processes which are deeply bifid or trifid at the apex; end view with three or four rays.

Staurastrum bifidum, Ralfs, *Annals of Nat. Hist.* v. 15. p. 151. t. 10. f. 3 (1845); *Trans. of Bot. Soc. of Edinburgh*, v. 2. p. 138. t. 13. (Not *Desmidium bifidum*, Ehr., or *Phycastrum bifidum*, Kützing.)
Goniocystis (Staurastrum) bifidum, Hassall, *Brit. Freshwater Algæ*, p. 355 (1845).

Dolgelley and Penzance, *J. R.* Near Bristol, *Mr. Broome*. Ambleside, *Mr. Sidebotham*. Craigendurroch and Glen Lin, Aberdeenshire, *Dr. Dickie*. Aberdeen, *Mr. P. Grant*. Reigate, *Mr. Jenner*.

Frond minute, smooth, scarcely constricted at the middle, truncate at the ends; segments with three or four elongated, straight, tapering processes, which are directed outwards, and consequently diverge from those of the other segment. In the front view each segment usually exhibits only two processes, the others being hidden behind them. The processes are usually trifid at the extremity, though sometimes merely bifid, in which latter case their diverging points are forked like the tail of a swallow: the same bifid appearance may occur in certain positions, even when three points are present. The end view has three or four rays, hyaline, stout at their base, and gradually tapering. In this view, as the frond is generally twisted, the rays of the lower segment may be faintly seen between those of the nearer one.

Sporangia have been gathered by Mr. Broome at Yate near Bristol; they are quadrate and spinous, and afford the only known example of a quadrate sporangium bearing spines. The emptied segments are easily detached; the spines are subulate and few.

Staurastrum brachiatum may always be distinguished from *S. paradoxum* and *S. tetracerum* by its smooth and divided processes.

Length of frond $\frac{1}{1111}$ of an inch; breadth $\frac{1}{1785}$; length of process from $\frac{1}{1633}$ to $\frac{1}{1165}$; length of sporangium from $\frac{1}{1020}$ to $\frac{1}{808}$; length of spine from $\frac{1}{2732}$ to $\frac{1}{2048}$.

Tab. XXIII. fig. 9. *a*. front view of frond; *b, c*. dividing fronds; *d*. end view of four-rayed variety; *e, f, g*. sporangia.

19. **S. læve** ——; frond smooth; segments with short processes forked at the apex and directed outwards; end view with three or four bipartite angles.

Pool near the outlet of Llyn Gwernan, Dolgelley, *J. R.*

Fronds very minute, smooth, deeply constricted at the middle, the constriction producing a wide triangular notch on each side. Segments externally lunate or somewhat cuneate, but in general the outer margin is slightly protuberant at the middle. Each angle terminated by a pair of short, stout, hyaline processes, which are slightly forked at the apex and directed outwards; but in the front view seldom more than three or four processes are seen at once. End view with three or four bipartite angles, their divisions divergent, subulate, generally appearing acute at the apex, but sometimes, especially if looked at obliquely, forked as in the front view.

Staurastrum læve, in the front view, agrees with *S. brachiatum* in its smooth frond and forked divergent processes; but the constriction at the middle is greater, the processes are shorter, and if an angle be towards the eye its geminate character will be detected without much difficulty. In the end view this species is unlike any other.

Length of frond $\frac{1}{1220}$ of an inch; breadth $\frac{1}{2127}$; breadth at constriction $\frac{1}{2500}$; length of process from $\frac{1}{3546}$ to $\frac{1}{3268}$.

Tab. XXIII. fig. 10. *a, b, c.* front views; *d.* end view; *e.* transverse view.

******* *Frond rough with puncta-like granules.*

† *End view with entire, rounded or truncate, angles or short rays.*

20. **S. alternans** (Bréb.); segments rough, narrow-oblong, and, from their twisted position, unequal in the front view; end view with the angles of one segment entire, and alternating with those of the other.

Staurastrum tricorne, Ralfs, *in Annals of Nat. Hist.* v. 15. p. 141. t. 11. f. 2 (1845); *Trans. of Bot. Soc. of Edinburgh*, v. 2. p. 141. t. 14. Jenner, *Fl. of Tunbridge Wells*, p. 194. (Not of Meneghini according to Brébisson.)

Goniocystis (Trigonocystis) hexaceros, Hassall, *Brit. Freshwater Algæ*, p. 352 (1845).

Staurastrum alternans, Brébisson, *in lit cum icone* (1846).

Dolgelley and Penzance, *J. R.* Barmouth, *Rev. T. Salwey.* Weston Bogs near Southampton; and several stations in Sussex, Surrey and Kent, *Mr. Jenner.* Near Bristol, *Mr. Thwaites.* Glen Lin, Aberdeenshire (alt. 1300 feet), *Dr. Dickie.* Near Aberdeen, *Mr. P. Grant.* Ambleside, *Mr. Sidebotham.*

Falaise, *Brébisson.* West Point, New York, *Bailey.*

Frond rough with minute pearly granules, which, except on the margin, appear like puncta; segments two or three times longer than broad, oblong, twisted, so that in the front view one of them appears shorter on one side, in consequence of the blending together of two of the angles. The end view is triangular, with concave sides and very obtuse entire angles. The angles of the lower segment are seen alternating with those of the upper.

I formerly described this plant as the *Staurastrum tricorne*, but that species in the front view is prolonged at the sides into short processes; I am not certain that the two are distinct, but in doubtful points I think it right to defer to M. de Brébisson's opinion.

Staurastrum alternans may be known from *S. dilatatum* and *S. punctulatum* by its unequal segments in the front view and alternating angles in the end one.

I have gathered the sporangia at Penzance; they are orbicular and furnished with spines forked at the apex.

Length of frond $\frac{1}{1037}$ of an inch; breadth $\frac{1}{1106}$; breadth at constriction $\frac{1}{3205}$.

Tab. XXI. fig. 7. *a.* front view; *b.* end view; *c.* transverse view.

21. **S. *punctulatum*** (Bréb.); segments rough with puncta-like granules, elliptic, equal; end view with broadly rounded angles and slightly concave sides.

Staurastrum punctulatum, Brébisson, *in lit cum icone* (1846).

Sussex, *Mr. Jenner.* Dolgelley, *J. R.*

Falaise, *Brébisson.*

Frond larger than that of *Staurastrum alternans*, rough with minute pearly granules which appear like puncta; segments twice as broad as long, elliptic; end view triangular with very blunt angles.

Staurastrum punctulatum may be distinguished from *S. alternans* by its equal and more turgid segments in the front view, and in an end one by its less concave sides. *S. rugulosum* agrees with it in size and partly in form, but in that species the pearly granules are larger and fewer, and at the angles appear like little spines.

Length of frond $\frac{1}{764}$ of an inch; breadth $\frac{1}{881}$; breadth at constriction $\frac{1}{2270}$.

Tab. XXII. fig. 1. *a.* front view; *b.* end view of empty frond; *c.* dividing fronds.

22. **S. *dilatatum*** (Ehr.); segments rough, fusiform, equal; end view with four short, broad, truncate and entire rays.

Staurastrum dilatatum, Ehr. *Infus.* p. 143. t. 10. f. 13 (1838). Menegh. *Synop. Desmid. in Linnæa* 1840, p. 156. Pritch. *Infus.* p. 184. Ralfs, *Annals of Nat. Hist.* v. 15. p. 143. t. 11. f. 5; *Trans. of Bot. Soc. of Edinburgh*, v. 2. p. 143. t. 14.

Phycastrum dilatatum, Kützing, *Phyc. Germ.* p. 138 (1845).

Goniocystis (Staurastrum) dilatatum, Hassall, *Brit. Freshwater Algæ*, p. 353 (1845).

Dolgelley and Penzance, *J. R.* Rusthall Common near Tunbridge Wells, and Rackham near Pulborough, Sussex, *Mr. Jenner.* Yate near Bristol, *Mr. Broome.* Ambleside, *Mr. Sidebotham.*

Germany, *Ehrenberg, Kützing.* Falaise, *Brébisson.* West Point, New York, *Bailey.*

Frond very minute, rough with puncta-like pearly granules, deeply constricted at the middle, the sinuses rounded; segments fusiform, two or three times broader than long, equal, obtuse at the sides, and not elongated into processes. In the end view, which is quadrangular, the sides are concave, and the angles form short, broad, truncate rays, on which the granules are arranged in transverse lines. Meneghini describes the rays as varying in number from three to five.

Staurastrum dilatatum differs from *S. alternans* in not being twisted; its rays also (in an end view) are more truncate.

Length of frond $\frac{1}{1201}$ of an inch; breadth $\frac{1}{1381}$; breadth at constriction $\frac{1}{3731}$.

Tab. XXI. fig. 8. *a.* front view of frond; *b.* end view; *c.* transverse view.

23. *S. margaritaceum* (Ehr.); segments rough tapering at the constriction, and having short lateral processes; end view with five or more short, narrow, obtuse rays.

Pentasterias margaritacea, Ehr. *Infus.* p. 144. t. 10. f. 15 (1838). Pritch. *Infus.* p. 185. f. 104.
Staurastrum margaritaceum, Menegh. *Synop. Desmid. in Linnæa* 1840, p. 227. Ralfs, *Annals of Nat. Hist.* v. 15. p. 157. t. 11. f. 7; *Trans. of Bot. Soc. of Edinburgh,* v. 2. p. 144. t. 14.
Phycastrum margaritaceum, Kützing, *Phycologia Germ.* p. 138 (1845).
Goniocystis (Pentasterias) margaritacea, Hassall, *Brit. Freshwater Algæ,* p. 356 (1845).

Dolgelley, *J. R.* Ashdown Forest, Sussex, *Mr. Jenner.* Stoke-Hill near Wells, *Mr. Thwaites.* Aberdeenshire, *Dr. Dickie* and *Mr. P. Grant.* Ambleside, *Mr. Sidebotham.*

Germany, *Ehrenberg.* Falaise, *Brébisson.* West Point, New York, *Bailey.*

Frond rough with minute pearly granules; in the front view the segments are convex at the ends and slightly attenuated at their junction, and on each side there is a short, linear, obtuse and entire process, which is generally somewhat incurved; occasionally these processes are mere conical projections, but they may always be perceived. The end view has from five to seven short, narrow, obtuse, marginal rays.

M. de Brébisson unites *S. margaritaceum* to *S. dilatatum,* but they appear to me to be distinct. In the front view of the latter the segments do not taper at their junction, are broader than long, and have no processes visible at the sides; in all which respects it differs from *S. margaritaceum.*

Length of frond $\frac{1}{1176}$ of an inch; breadth, including processes, $\frac{1}{1000}$; breadth, excluding processes, $\frac{1}{2358}$; breadth at constriction $\frac{1}{3378}$.

Tab. XXI. fig. 9. *a, b.* front views; *c, d, e.* end views.

24. *S. tricorne* (Bréb.); frond rough with puncta-like granules; segments tapering at each side into a short, blunt, mostly entire process; end view with three or four blunt angles.

β. Processes terminated by minute spines.

Binatella tricornis, Brébisson, *Alg. Fal.* p. 57. t. 8 (1835).
Desmidium hexaceros, Ehr. *Infus.* p. 141. t. 10. f. 10 (1838).
Staurastrum tricorne, Meneghini, *Synop. Desmid. in Linnæa* 1840, p. 225. Brébisson, *in lit. cum icone.*

Phycastrum tricorne, Kützing, *Phycologia Germanica*, p. 137 (1845).

Dolgelley and Penzance, *J. R.* Ambleside, *Mr. Sidebotham.* Cross-in-Hand, Sussex, *Mr. Jenner.*

β. Penzance, *J. R.* Near Pulborough, Sussex, *Mr. Jenner.*

Falaise, *Brébisson.* Germany, *Ehrenberg, Kützing.*

Frond about as large as that of *Staurastrum alternans*, deeply constricted at the middle; segments in the front view somewhat fusiform; in an end one showing three or four blunt, generally entire angles.

The segments are frequently more or less twisted, in which case this plant bears a close resemblance to *Staurastrum alternans*; but I retain the species in deference to the authority of M. de Brébisson; the tapering of the segments at their sides is however its chief distinction.

The sporangia, which I have gathered at Penzance, are orbicular, and furnished with spines divided at the apex.

Length of frond $\frac{1}{1275}$ of an inch; breadth $\frac{1}{948}$; breadth at constriction $\frac{1}{3731}$; diameter of sporangium $\frac{1}{680}$; length of spines $\frac{1}{1457}$.

β. Length of frond from $\frac{1}{1000}$ to $\frac{1}{972}$; breadth from $\frac{1}{796}$ to $\frac{1}{697}$; breadth at constriction $\frac{1}{3267}$; diameter of sporangium $\frac{1}{909}$; length of spines $\frac{1}{1633}$.

Tab. XXII. fig. 11. *a, b.* front views of frond; *c.* end view.

Tab. XXXIV. fig. 8. *a.* sporangium; *b.* front view of var. β; *c.* end view; *d.* sporangium.

†† *Angles terminated by minute spines or tapering into slender processes*.*

25. *S. polymorphum* (Bréb.); segments rough with minute granules, having on each side a short process tipped with spines; end view three- to six-rayed.

Staurastrum polymorphum, Brébisson, *in lit.* (1846).

Dolgelley, *J. R.* Reigate, Surrey; and near Pulborough, Sussex, *Mr. Jenner.* Falaise, *Brébisson.*

Frond much smaller than that of *Staurastrum gracile*, deeply constricted at the middle; segments irregular in form, but generally broader than long. Each side terminates in a short truncate projection or process, which is scarcely longer than broad, and tipped by three or four distinct, diverging spines; frequently there are also a few inconspicuous scattered spines on the segment itself. The number of the angles or rays in an end view varies from three to six, but the four-rayed form is the most abundant. The size of the frond is proportionate to the number of these rays.

The sporangia, which I have gathered at Dolgelley, are orbicular; their spines are few and forked at the apex. I have seen three-rayed fronds coupled with four-rayed ones.

---

* A variety of *Staurastrum tricorne* has the angles tipped by minute spines, and occasionally *Staurastrum cyrtocerum* is destitute of spines, excepting those terminating the processes; both forms would therefore be referred to this section.

The segments are very variable in form and often resemble those of *Staurastrum tricorne* and *S. margaritaceum*, but in these species the processes are never spinous. *S. polymorphum* is smaller and less spinous than *S. asperum*.

Length of frond $\frac{1}{1000}$ of an inch; breadth $\frac{1}{1157}$; breadth at constriction $\frac{1}{2500}$; diameter of sporangium $\frac{1}{1000}$; length of spine of sporangium $\frac{1}{1106}$.

Tab. XXII. fig. 9. *a, b, c, d.* front views; *e, f, g.* dividing fronds; *h.* end view of three-rayed segment; *i.* transverse view of four-rayed segment; *k.* end view with endochrome; *l.* end view of empty frond.

Tab. XXXIV. fig. 6. *a.* front view of variety; *b.* end view.

26. **S. gracile** (Ralfs); **segments rough, elongated on each side into a slender process which is terminated by minute spines; end view triradiate.**

Euastrum, No. 13, Bailey, *Amer. Bacil. in Amer. Journ. of Science and Arts*, v. 41. p. 296. t. 1. f. 2–5 (1841).
Staurastrum gracile, Ralfs, *Annals of Nat. Hist.* v. 15. p. 155. t. 11. f. 3 (1845); *Trans. of Bot. Soc. of Edinburgh*, v. 2. p. 142. t. 14.
Goniocystis (Trigonocystis) gracilis, Hassall, *Brit. Freshwater Alg.* p. 352 (1845).

Dolgelley and Penzance, *J. R.* Ambleside, *Mr. Sidebotham.* Aberdeen, *Mr. P. Grant.* Manchester, *Mr. Williamson.*

New York and New England, *Bailey.* Falaise, *Brébisson.*

Frond rough with minute granules, deeply constricted at the middle; segments two or three times longer than broad and tapering, on each side, into a slender, straight and colourless process, which is apparently terminated by three minute points: Mr. Jenner however informs me that the processes really terminate in four points, which are visible only when a process presents its extremity to the observer. The granules are arranged in transverse lines on the processes, and are here more conspicuous than they are on the segment. The end view is triradiate; the colouring matter is restricted to the centre and forms three rays, which are frequently bifid.

Staurastrum gracile differs from *S. tricorne* in its elongated processes terminated by minute points. The end view somewhat resembles that of *S. paradoxum*, but the latter species is easily known by the divergent processes of its front view and its smaller size.

Length of frond from $\frac{1}{773}$ to $\frac{1}{539}$ of an inch; breadth from $\frac{1}{348}$ to $\frac{1}{272}$; breadth at constriction $\frac{1}{3105}$.

Tab. XXII. fig. 12. *a.* front view; *b, c.* end views; *d.* transverse view.

27. **S. Arachne** (Ralfs); **segments rough with minute granules, suborbicular, with elongated, slender, incurved processes; end view with five linear rays.**

Staurastrum Arachne, Ralfs, *Annals of Nat. Hist.* v. 15. p. 157. t. 11. f. 6 (1845); *Trans. of Bot. Soc. of Edinburgh*, v. 2. p. 143. t. 14.

Goniocystis (*Pentasterias*) *arachnis,* Hassall, *Brit. Freshwater Alg.* p. 355 (1845).

Dolgelley, very rare, *J. R.* Aberdeen, *Mr. P. Grant.*
Falaise, *Brébisson.*

Frond minute, deeply constricted at the middle; segments about as long as broad, having on each side an elongated process, which is hyaline and incurved, and on account of its minute granules appears transversely striated. When the frond is viewed obliquely, so that three or more of the long curved processes are seen at once, its resemblance to an insect is considerable. The end view is circular with five slender rays.

This plant is remarkable for its slender processes, which will easily distinguish it from *Staurastrum margaritaceum.* It cannot be a five-rayed variety of *S. gracile* or *S. polymorphum,* for its rays are longer, more slender, remarkably incurved and also entire at the extremity.

Length of frond $\frac{1}{1020}$ of an inch; breadth, excluding processes, $\frac{1}{2040}$; including processes, $\frac{1}{652}$; length of process $\frac{1}{1858}$; breadth at constriction $\frac{1}{2732}$.

Tab. XXIII. fig. 6. *a.* front view; *b.* end view.

28. *S. tetracerum* (Kützing); frond rough; front view with four slender diverging processes which are entire at the apex; end view compressed, with a process at each extremity.

Micrasterias tetracera, Kützing, *Synopsis Diatom. in Linnæa* 1833, p. 602. f. 83, 84.
Micrasterias tricera, Kützing, *l. c.* p. 602. f. 85.
Staurastrum paradoxum, Ehr. *Infus.* p. 143. t. 10. f. 14 (1838). Pritch. *Infus.* p. 185. f. 102, 103.
Staurastrum tetracerum, Ralfs, *Annals of Nat. Hist.* v. 15. p. 150. t. 10. f. 1 (1845); *Trans. of Bot. Soc. of Edinburgh,* v. 2. p. 137. t. 13.
Goniocystis (*Staurastrum?*) *tetracerum,* Hassall, *Brit. Freshwater Alg.* p. 354 (1845).

Dolgelley and Penzance, *J. R.* Yate near Bristol, *Mr. Broome.* Aberdeenshire, *Mr. P. Grant.* Ambleside, *Mr. Sidebotham.* Midhurst, Battle, and Cross-in-Hand, Sussex, *Mr. Jenner.*

Germany, *Kützing, Ehrenberg.* Falaise, *Brébisson.* United States, *Bailey.*

Frond very minute; front view nearly square, the angles elongated into straight, slender processes which diverge from each other; frequently however a segment may be so twisted, that one of its processes is situated behind its companion, and is not seen unless carefully looked for; in this case the frond seems to have only three processes in a front view. The end view is much compressed and terminated both ways by an elongated process. The frond is rough with minute puncta-like granules, which form transverse lines on the processes and give them a jointed appearance. The colouring matter is very pale.

Ehrenberg and Meneghini unite the *Micrasterias tetracera* of Kützing to the *Staurastrum paradoxum*, Meyen; but whilst the latter plant has four processes at each end, this has only two, which, if I am correct in my view of the following species, differ also by having entire extremities.

Length of frond $\frac{1}{2703}$ of an inch; length of process $\frac{1}{3030}$; breadth of frond $\frac{1}{1785}$.

Tab. XXIII. fig. 7. *a, b, c.* front views; *d.* empty frond; *e.* frond with one segment not fully developed; *f.* end view.

29. **S. paradoxum** (Meyen); frond rough; front view with elongated diverging processes which are minutely trifid at the apex; end view quadrangular or sometimes triangular.

β. End view triradiate.

Staurastrum paradoxum, Meyen, *Nov. Act. Leop. Holm.* v. 14. p. 43. f. 37, 38 (1828). Meneghini, *Synop. Desmid. in Linnæa* 1840, p. 227. Ralfs, *Annals of Nat. Hist.* v. 15. p. 151. t. 10. f. 2; *Trans. of Bot. Soc. of Edinburgh*, v. 2. p. 137. t. 13.

Micrasterias Staurastrum, Kützing, *Synop. Diatom. in Linnæa* 1833, p. 599.

Phycastrum paradoxum, Kützing, *Phycologia Germanica*, p. 138 (1845).

Goniocystis (Staurastrum) paradoxum, Hassall, *Brit. Freshwater Alg.* p. 354 (1845).

Dolgelley and Penzance, *J. R.* Near Bristol, *Mr. Thwaites* and *Mr. Broome.* Ambleside, *Mr. Sidebotham.* Banffshire and Aberdeenshire, *Mr. P. Grant.* Near Pulborough, Cross-in-Hand, &c., Sussex, *Mr. Jenner.*

Germany, *Meyen, Kützing.* Falaise, *Brébisson.* West Point, New York, *Bailey.*

Frond very minute, generally much constricted at the middle, the ends truncate; each segment has, generally, four elongated processes diverging from those of the other segment. The end view is sometimes triradiate, but usually quadriradiate; I have also seen specimens in which one segment had four and the other only three rays. In the front view, as the frond generally rests on one side, each segment appears to have only two processes, the other two being covered by them; but sometimes only one process is thus hidden, and occasionally all four may be seen at the same time. The processes are elongated, colourless, and, being rough with minute granules arranged in transverse lines, have a jointed appearance; each terminates in three minute spines. In the newly-formed segment the processes are mere conical projections.

When only two processes are seen at each end, in the front view, this species is very like the preceding; but its frond is more constricted, larger and of a deeper green, and its processes are thicker and not entire at the apex.

Length of frond $\frac{1}{941}$ of an inch; breadth $\frac{1}{1165}$; breadth at constriction $\frac{1}{2923}$; length of process from $\frac{1}{1362}$ to $\frac{1}{1256}$.

Tab. XXIII. fig. 8. *a, b, c.* front views of mature fronds; *d, e.* end views; *f.* immature frond.

******** *Frond spinous or rough with spine-like granules which are incrassated, emarginate, or divided**.

† *Spines minute, inconspicuous and granule-like.*

30. **S. cyrtocerum** (Bréb.); frond rough with minute granules; segments in the front view somewhat triangular with short incurved processes.

Euastrum ——, Bailey, *American Journ. of Science and Arts*, v. 41. t. 1. f. 6 (1841).
Staurastrum cyrtocerum, Brébisson, *in lit. cum icone* (1846).

Dolgelley and Penzance, *J. R.* Near Bristol, *Mr. Broome*. Piltdown Common and near Pulborough, Sussex, *Mr. Jenner*.

United States, *Bailey*. Falaise, *Brébisson*.

Frond rather larger than that of *Staurastrum tricorne*, deeply constricted at the middle, rough with minute spine-like granules, which on the processes are arranged in transverse lines. The segments are truncate and taper towards their junction, so that the constriction forms a broad notch on each side. The spines on the outer margin are larger than those on the rest of the segment, and frequently are obscurely notched. The end view has three blunt angles, much like those of *Staurastrum tricorne*, and usually a little curved.

The sporangia are orbicular and their spines slightly forked at the apex.

Staurastrum cyrtocerum is smaller than *S. aculeatum* and *S. controversum*, and its spines are far less conspicuous. It is best distinguished by its converging processes.

Length of frond $\frac{1}{800}$ of an inch; breadth $\frac{1}{800}$; breadth at constriction $\frac{1}{2747}$.

Tab. XXII. fig. 10. *a*. front view; *b*. end view.

31. **S. asperum** (Bréb.); segments elliptic or somewhat cuneiform, rough with minute spines, which on the outer margin are usually dilated at the end or forked.

a. Angles in end view rounded.

β. *proboscideum*, Bréb.; angles in end view prolonged into short rays terminated by minute spines.

Staurastrum asperum, Brébisson, *in lit. cum icone* (1846).

Penzance, *J. R.* Yate near Bristol, *Mr. Broome* and *Mr. Thwaites*. Near Pulborough, Sussex, *Mr. Jenner*.

Falaise, *Brébisson*.

Frond comparatively large, rough with minute acute granules or spines; segments slightly tapering on the inner side, so that their figure is somewhat cuneate, and the constriction forms a wide notch on each side. The outer

---

* *Staurastrum polymorphum* has occasionally a few scattered spines, but they are simple.

margin is flattened, and its spines are usually more evident than the rest, and are also in general slightly forked. In a. the lateral margins are broadly rounded, but in β. they extend into a process about as long as broad and tipped with a few acute spines, which are larger than those on the segment itself. The end view, in which the sides are nearly straight, has the angles in a. rounded, but in β. terminated by short rays.

The sporangia, which I have gathered at Penzance, are orbicular, and their spines are twice branched at the apex.

The form of the segments distinguishes *Staurastrum asperum* from all states of *S. muricatum*. *S. punctulatum* is smaller, and its granules are more like puncta than spines. In *S. rugulosum* the segments in the front view are elliptic, not tapering at the junction, and its spine-like granules are confined to the angles.

Length of frond $\frac{1}{555}$ of an inch; breadth $\frac{1}{615}$; breadth at constriction $\frac{1}{2403}$; diameter of sporangium $\frac{1}{519}$; length of spines $\frac{1}{2040}$.

Tab. XXII. fig. 6. a. front view; b. end view.

Tab. XXIII. fig. 12. a. sporangium; b. front view of variety β; c. end view.

†† *End view acutely triangular; segments, in the front view, with a forked spine on each side, otherwise smooth.*

32. **S. Avicula** (Bréb.); segments with a forked spine on each side; each angle, in end view, terminated by a mucro-like spine.

Staurastrum Avicula, Brébisson, *in lit. cum icone* (1846).

Penzance, very rare, *J. R.*

Falaise, *Brébisson*.

Frond very minute, scarcely rough, the constriction producing wide triangular notches; segments having on each side a spine forked like the tail of a swallow. End view with three slightly inflated angles or lobes, which are tipped by a spine.

The forked lateral spines of the front view mark the species.

Length of frond $\frac{1}{907}$ of an inch; breadth $\frac{1}{948}$; breadth at constriction $\frac{1}{2403}$; length of spine $\frac{1}{4098}$.

Tab. XXIII. fig. 11. a. front view; b. end view.

††† *Lobes in end view broad, emarginate or bipartite.*

33. **S. enorme** ——; frond irregular or quadrate, spinous; end view three- or four-lobed; lobes broad, more or less emarginate or bifid, and terminated by spines which are either simple or branched.

Dolgelley, *J. R.*

Frond very irregular and variable in form. Sometimes the front view differs but little from the end one, usually however there is a slight constriction or sinus at the junction of the segments, but I have never observed any difference in the endochrome at that part. The spines, which are almost confined to the angles, are irregular, some simple and some branched. The end view has three or four broad and very irregular lobes; these are spinous and more or less emarginate, and frequently one lobe is much broader and more spinous than the others. The spines on each lobe form two groups separated by the notch; they vary much in size, and are either simple and subulate or else forked; sometimes the forked spines are again divided at the apex.

Staurastrum enorme is by far the least symmetrical plant in this family, especially in its front view, and it is very difficult to trace any division into segments. In the end view the broadly emarginate lobes, which, exclusive of the spines, are truncate, afford a good specific distinction between it and any other species.

Length of frond $\frac{1}{601}$ of an inch; breadth of lobe in the end view $\frac{1}{1497}$.
Tab. XXXIII. fig. 11. *a, b, c.* front views; *d, e.* end views.

†††† *Spines numerous, conspicuous; angles in end view either rounded, acute, or extended into processes.*

34. **S. *spongiosum*** (Bréb.); segments semiorbicular, spinulose; spines forked; end view triangular, bordered with forked spines; angles rounded.

Staurastrum spongiosum, Brébisson, Meneghini, *Synop. Desmid. in Linnæa* 1840, p. 229. Brébisson, *in lit. cum icone.*
Desmidium ramosum, Ehrenberg, *Verbreitung und Einfluss des mikroskopischen Lebens in Süd- und Nord-Amerika*, t. 4. f. 21? (1843).

Penzance and Dolgelley, rare, *J. R.* Ashdown Forest, Sussex, *Mr. Jenner.*

Falaise, *Brébisson.* West Point, New York, *Bailey.*

Frond large, about the size of that of *Staurastrum muricatum*, thickly covered with short spines, which are forked at the apex; segments semiorbicular, having on each side one spine more conspicuous and more forked than the rest. End view triangular, fringed with short notched spines; the sides are slightly convex and the angles rounded.

Staurastrum spongiosum is a very distinct species. In the front view it has some resemblance to *Xanthidium armatum*, but is much smaller.

Length of frond from $\frac{1}{506}$ to $\frac{1}{418}$ of an inch; breadth from $\frac{1}{523}$ to $\frac{1}{476}$; breadth at constriction from $\frac{1}{833}$ to $\frac{1}{738}$.
Tab. XXIII. fig. 4. *a, b.* front views; *c.* end view.

35. **S. *controversum*** (Bréb.); frond spinulose; segments with a short process on each side terminated by minute spines; end view with three or four distorted rays.

Staurastrum controversum, Brébisson, Meneghini, *Synop. Desmid. in Linnæa* 1840, p. 228. Bréb. *in lit. cum icone.*

Staurastrum aculeatum, Ralfs, *Annals of Nat. Hist.* v. 15. p. 156. t. 11. f. 4 (1845); *Trans. of Bot. Soc. of Edinburgh,* v. 2. p. 142. t. 14. Jenner, *Fl. of Tunbridge Wells,* p. 194.
Goniocystis (Trigonocystis?) aculeata, Hassall, *Br. Freshwater Alg.* p. 353 (1845).

Dolgelley and Penzance, *J. R.* Reigate and Woking Common, Surrey; Cross-in-Hand, Piltdown Common, &c., Sussex, *Mr. Jenner.* Yate near Bristol, *Mr. Broome.* Ambleside, *Mr. Sidebotham.* Aberdeen, *Mr. P. Grant.*

Falaise, *Brébisson.*

Frond deeply constricted at the middle; segments broader than long, elliptic or fusiform, often irregular, but usually turgid on the inner side, laterally tapering into a short process, which is terminated by minute spines and generally curved or distorted. The end view is triangular or quadrangular, the angles terminating in short, truncate, curved processes, which, as in the front view, are tipped by minute subulate spines. In both views the fronds show numerous conspicuous spines which are either subulate or notched at the end.

I formerly considered this plant to be the *Staurastrum aculeatum,* and I still doubt whether it may not be a variety of that species, which scarcely differs except in its processes, which are always straight.

Length of frond $\frac{1}{972}$ of an inch; breadth from $\frac{1}{880}$ to $\frac{1}{709}$; breadth at constriction $\frac{1}{3731}$.

Tab. XXIII. fig. 3. *a, b, c.* front views; *d.* end view of three-rayed form; *e.* transverse view; *f.* view of four-rayed variety; *g.* transverse view; *h.* sporangium from a drawing by Brébisson.

36. **S. aculeatum** (Ehr.); frond spinulose; segments with a short process on each side terminated by minute spines; end view with three to five straight rays terminated by spines.

Desmidium aculeatum, Ehrenberg, *Infus.* p. 143. t. 10. f. 12 (1838). Pritchard, *Infus.* p. 184.
Staurastrum aculeatum, Meneghini, *Synop. Desmid. in Linnæa* 1840, p. 226.
Phycastrum aculeatum, Kützing, *Phycologia Germanica,* p. 138 (1845).

Weston Bogs near Southampton; and near Pulborough, Sussex, *Mr. Jenner.* Dolgelley, *J. R.* Ambleside, *Mr. Sidebotham.*

Germany, *Ehrenberg.* Falaise, *Brébisson.*

Frond about as large as that of *Staurastrum controversum,* deeply constricted at the middle, the constriction producing a broad triangular notch on each side; segments broader than long, somewhat fusiform, thickly spinulose, the spines simple or notched at the apex. The segments taper on each side into a short process tipped by three or four spines. The end view has from three to five short straight rays terminated by spines.

Length of frond $\frac{1}{666}$ of an inch; breadth $\frac{1}{500}$; breadth at constriction $\frac{1}{1666}$.

Tab. XXIII. fig. 2. *a.* front view; *b.* end view of three-rayed form; *c.* end view of four-rayed variety.

37. *S. spinosum* (Bréb.); segments elliptic, furnished with a few bifid spines; lateral spines solitary, larger and more forked; end view triangular with two or three spines on each side and one terminating each angle.

Staurastrum spinosum, Brébisson, *in lit. cum icone* (1847).

Dolgelley and Penzance, *J. R.* Ambleside, *Mr. Sidebotham.* Reigate, Surrey, *Mr. Jenner.*

Falaise, *Brébisson.*

Frond deeply constricted at the middle; segments subelliptic, their outer margin usually more turgid than the inner; on each side is a conspicuous sessile spine, forked like the tail of a swallow, and on the outer margin are a few smaller spines, which are usually forked at the end, though sometimes simple. End view triangular, each angle terminated by a spine, which appears simple or forked according to its position when viewed; sides with two or three spines, which are sometimes simple and sometimes forked, and on the upper surface are six other spines, one opposite to each lateral spine.

The sporangia of *Staurastrum spinosum* were first gathered by Mr. Broome near Bristol, and I have since found a few of them at Dolgelley. They are orbicular, and have numerous elongated spines which are divided at the apex.

M. de Brébisson considers this plant identical with the *Xanthidium furcatum*, Ehr.

Length of frond $\frac{1}{859}$ of an inch; breadth $\frac{1}{907}$; breadth at constriction $\frac{1}{3267}$; diameter of sporangium $\frac{1}{641}$; length of spine of sporangium $\frac{1}{1408}$.

Tab. XXII. fig. 8. *a.* front view of frond; *b.* end view; *c, d.* sporangia.

38. *S. vestitum* ——; frond rough with minute emarginate spines; segments fusiform; end view triradiate, each side having two spines, short, slender, and often accompanied by other smaller ones.

Dolgelley, *J. R.* Near Pulborough, Sussex, *Mr. Jenner.*

Frond broader than long, about as large as that of *Staurastrum gracile*, deeply constricted at the middle, the constriction forming a triangular notch on each side; segments somewhat fusiform, turgid on the inner margin, tapering at each side into a short process which is tipped by three or four minute spines; outer margin fringed with minute emarginate spines or tubercles. End view triradiate, showing also two slender forked spines at the middle of each side, whilst the margin is often fringed with other spines, which are smaller and either simple or notched. The rays are elongate, rough with minute granules, and terminated by minute subulate spines.

In the end view *Staurastrum vestitum* resembles *S. gracile* in figure, but that species is not spinous. The rays in the end view are more slender than those of *S. aculeatum* or *S. controversum*: but the most distinctive character of this species is the presence of a pair of slender forked spines at the middle of each margin.

Length of frond $\frac{1}{625}$ of an inch; breadth $\frac{1}{384}$; breadth at constriction $\frac{1}{3205}$.

Tab. XXIII. fig. 1. *a.* front view; *b, c.* end views with endochrome; *d.* end view of empty frond; *e.* transverse view.

12. DIDYMOCLADON, n. g.

Frond simple, constricted at the middle, angular, each angle having two processes, one lateral, and in front view nearly parallel to the adjacent one of the other segment, the other superior and divergent.

The frond is simple, constricted at the middle; segments angular, as shown by the end view, each angle furnished with two elongated processes, the inferior lateral and nearly parallel to the similar one of the other segment, but the superior directed obliquely outwards and divergent.

This genus is closely allied to *Staurastrum*, but has one remarkable differential character. Each angle of the segments gives rise to two processes, one beneath the other, whereas in *Staurastrum* the process, when present, is solitary from each angle.

D. furcigerus, Bréb.

    a. End view triangular.

    β. End view quadrangular.

Staurastrum furcigerum, Bréb., *Meneghini, Synop. Desmid. in Linnæa* 1840, p. 226; *in lit. cum icone.*
Phycastrum furcigerum, Kützing, *Phy. Germ.* p. 138 (1845).

Dolgelley, *J. R.* Brookhouse Moss near Congleton, Cheshire; and near Ambleside, *Mr. Sidebotham.* Rochdale, *Mr. Coates.* Brightling near Burwash, Sussex; and Weston Bogs near Southampton, *Mr. Jenner.*

Falaise, *Brébisson.* Germany, *Kützing.*

Frond comparatively large, rough with pearly granules, which being arranged on the processes in transverse lines, produce a crenate appearance on their margins. In the front view the processes are elongated, stout, tapering, bifid at the apex. The lateral processes of the one segment are usually parallel to those of the other; in Rochdale specimens however they are slightly curved outwards. The end view is triangular or quadrangular, and the elongated angles appear either acute and simple, or bifid, according as one or two forks of the processes may be visible.

Length of frond, including processes, $\frac{1}{333}$ of an inch; length, excluding processes, $\frac{1}{547}$; breadth, including processes, $\frac{1}{357}$; breadth, excluding processes, $\frac{1}{555}$; breadth at constriction $\frac{1}{1666}$.

Tab. XXXIII. fig. 12. *a, b.* front views; *c.* end view of *a*; *d.* transverse view of β.

13. TETMEMORUS, *Ralfs.*

Frond simple, elongated, straight, cylindrical or fusiform, slightly constricted at the middle; *segments* emarginate at the end, but otherwise quite entire.

The frond is elongated, as in *Penium*, from which however this genus may be distinguished by the emarginate ends; the same character and the elongated frond will separate it from *Cosmarium*.

From *Euastrum*, with which it agrees in the emarginate extremities, it differs in being cylindrical or nearly so, and in the segments being neither lobed nor sinuated; the fronds are also free from inflated protuberances.

Sporangia have been gathered of two species.

1. *T. Brebissonii* (Menegh.); frond in the front view with parallel sides, but in the lateral view fusiform; the ends without any projecting processes; puncta in longitudinal lines.

β. *turgidus*; larger, constriction greater; segments somewhat inflated.

Closterium Brebissonii, Menegh. *Syn. Desmid. in Linnæa* 1840, p. 236.
Closterium (sp. 9), Bailey, *Amer. Bacil. in Amer. Journal of Science and Arts,* vol. 41. no. 2. t. 1. f. 38 (1841).
Tetmemorus Brebissonii, Ralfs, *Ann. of Nat. Hist.* v. 14. t. 8. f. 1 (1844); *Trans. of Bot. Soc. of Edinburgh*, v. 2. p. 133. t. 12. Jenner, *Fl. of Tunbridge Wells*, p. 198. Hassall, *Br. Freshwater Alg.* p. 377.
Closterium monile, Kützing, *Phycol. Germ.* p. 132 (1845).

Dolgelley; Carmarthen; Penzance, *J. R.* Sussex; Kent; Surrey and Hants, *Mr. Jenner.* Many stations in Aberdeenshire (alt. 50—3700 feet), *Mr. P. Grant* and *Dr. Dickie.* Banffshire, *Mr. P. Grant.* Ambleside, *Mr. Sidebotham.*

β. Ashdown Forest, Sussex, *Mr. Jenner.*

Falaise, *Brébisson.* Germany, *Kützing.* Maine to Virginia, *Bailey.*

Frond four to six times longer than broad, with a slight constriction at the middle. The front view shows the lateral margins of the segments nearly straight, and their ends rounded and emarginate. The lateral view is more constricted at the middle and the segments are attenuated towards the end.

The endochrome is dark green, and there is a series of large vesicles down the middle in a single row, either straight or with some irregularity. The frond, when empty, is found to be minutely punctate, the puncta being arranged in longitudinal rows.

The var. β. somewhat approaches in form to *Tetmemorus granulatus*, but is more constricted at the middle, and the puncta form longitudinal lines.

Tetmemorus Brebissonii differs from *T. granulatus* in having its front and lateral views unlike each other, and its puncta arranged in longitudinal lines; it is also generally smaller.

Length of immature frond $\frac{1}{384}$ of an inch; breadth $\frac{1}{1385}$; breadth at constriction $\frac{1}{1666}$; length of mature frond $\frac{1}{142}$; breadth $\frac{1}{704}$; breadth at constriction $\frac{1}{839}$.

Tab. XXIV. fig. 1. *a.* front view; *b.* side view; *c.* empty frond; *d, e.* front and side views of var. β; *f.* young frond.

2. **T. lævis** (Kützing); frond in the front view somewhat tapering, with truncate ends; lateral view fusiform; puncta none or very indistinct.

Tetmemorus granulatus (in part), Ralfs, *Annals of Natural Hist.* v. 14. t. 8. f. 2. *d, e, f, g* (1844); *Trans. of Bot. Soc. of Edinburgh*, v. 2. t. 12.
Closterium læve, Kützing, *Phycologia Germanica*, p. 132 (1845).

Dolgelley, *J. R.* Aberdeen, *Dr. Dickie.* Ashdown Forest and near Tunbridge Wells, Sussex, *Mr. Jenner.*

Germany, *Kützing.*

Frond four to six times longer than broad, constricted at the middle, tapering with straight sides; ends truncate, with a hyaline lip, which is inconspicuous, and sometimes absent.

The puncta are so faint, that even when using the higher powers of Mr. Ross's achromatic microscope I have doubted their existence; but Mr. Jenner and Mr. Ross assure me that they are scattered, as in *Tetmemorus granulatus.*

I have gathered the sporangia of this species for three successive years near Dolgelley, forming a mucous stratum on the moist soil; I have also seen them mixed with Desmidieæ sent from Aberdeen by Dr. Dickie.

After coupling, the segments of the fronds are separated by the formation of a large, quadrate, central cell, in which all the contents of both fronds are collected, the empty segments being loosely attached to its corners. The endochrome at first fills the cell, large starch globules being scattered throughout the minutely granular substance; but at length it becomes a dense, round, homogeneous body of a dark green colour, which finally changes to an olive-brown. In this stage the segments of the original fronds fall off, and leave the quadrate cell inclosing the sporangium. In the front view, as stated above, the cell is nearly square; the sides are concave, and the angles rounded and slightly produced. A lateral view shows that the cell and sporangium are both compressed.

In the present plant the process of forming the sporangium is interesting, as it exhibits a striking similarity to the change during the formation of similar bodies in *Staurocarpus* among the Conjugatæ. In *Staurocarpus*, after conjugation, a subquadrate cell is formed, within which the endochrome is collected. The latter is at first of the same figure as the cell, but in one species at least is at length condensed into a compact globular body, and in every species the cell with the contained sporangium finally separates from the filaments with which it is connected. In this separate state I can discover no character by which to distinguish the sporangium of *Tetmemorus* from one belonging to a species of *Staurocarpus*.

This plant is in size intermediate between *Tetmemorus Brebissonii* and *T. granulatus*. In figure it agrees with the former; like the latter it has a hyaline lip, but one far less conspicuous. This character induced me, in my paper read before the Botanical Society of Edinburgh, to refer its sporangia to that species, and I did not discover my error until I had seen the true sporangia of *T. granulatus*.

Tetmemorus lævis is distinguished from *T. Brebissonii* either by the absence of puncta or by their scattered state when visible; its fronds also taper more in the front view. From *T. granulatus* it differs in being more constricted at the middle, and in having the front and lateral views dissimilar, irrespective of the terminal notch.

Length of frond from $\frac{1}{374}$ to $\frac{1}{336}$ of an inch; breadth from $\frac{1}{1244}$ to $\frac{1}{1073}$; breadth at constriction from $\frac{1}{1322}$ to $\frac{1}{1147}$; length of side of quadrate sporangium $\frac{1}{447}$.

Tab. XXIV. fig. 3. *a*. front view; *b*. side view; *c*. empty frond; *d*, *e*, *f*. different stages of sporangium; *g*. side view of sporangium.

3. *T. granulatus* (Bréb.); frond fusiform both in the front and lateral views, and ending in a colourless projecting lip-like process.

Closterium granulatum, Brébisson, *Menegh. Syn. Desmid. in Linnæa* 1840, p. 236. Kützing, *Phy. Germ.* p. 132.
Tetmemorus granulatus, Ralfs, *Annals of Nat. Hist.* v. 14. t. 8. f. 2 (1844); *Trans. of Bot. Soc. of Edinburgh*, p. 134. t. 12. Jenner, *Fl. of Tunbridge Wells*, p. 198. Hassall, *Br. Freshwater Algæ*, p. 378.

Carnarvon; Dolgelley; Tal Sarn near Lampeter, and Penzance, *J.R.* Sussex, Surrey, Kent and Hampshire, *Mr. Jenner.* Kerry, *Mr. Andrews.* Hanham near Bristol, *Mr. Thwaites.* Charlton Fields, Manchester; and Ambleside, Westmoreland, *Mr. Sidebotham.* Aberdeenshire (alt. 1600 to 2455 feet), *Dr. Dickie.* Aberdeenshire and Banffshire (alt. 50 to 1600 feet), *Mr. P. Grant.*

Falaise, *Brébisson.* Germany, *Kützing.*

Frond fusiform, about six times as long as broad, and very slightly constricted at the middle. The extremities always have a colourless projecting lip-like process, which extends beyond the notch. The colouring is dark green, and a few large vesicles are arranged in a longitudinal row down the middle.

The empty frond is minutely punctated; the puncta generally form one or two transverse lines in each segment near the central constriction; in the other parts of the frond they are not in rows, but scattered.

Under a low power of the microscope this species much resembles the preceding; but it may always be distinguished by the front and lateral views being similar, and only differing in the terminal notch, which is not seen in the lateral view. There is also beyond the notch a remarkable lip-like projection, which is wanting in *Tetmemorus Brebissonii*, and the puncta seen in its empty frond are not arranged in longitudinal rows.

Tetmemorus granulatus differs from *T. lævis* in its less constricted middle and its larger size. Its puncta too are usually very distinct, and the discovery of their respective sporangia leaves no doubt that these plants are distinct.

Its sporangia have been gathered by Mr. Jenner in Sussex, and by myself near Dolgelley. They differ considerably from those of *T. lævis*; they are orbicular, and are not inclosed in a quadrate cell, but have the empty segments of the conjugating fronds loosely attached by an imperceptible membrane; the margin of the sporangium is finely striated, a character which I have not noticed in *Tetmemorus lævis*.

Length of frond from $\frac{1}{133}$ to $\frac{1}{130}$ of an inch; breadth $\frac{1}{649}$; breadth at constriction from $\frac{1}{736}$ to $\frac{1}{685}$; diameter of sporangium from $\frac{1}{365}$ to $\frac{1}{349}$.

Tab. XXIV. fig. 2. *a*. front view; *b*. side view; *c*. empty frond.

14. PENIUM, *Bréb*. (in lit.)

Frond simple, elongated, straight, not, or but slightly, constricted; *segments* entire.

The frond is cylindrical or elliptic and straight, with its opposite margins similar, its constriction none or very slight, and the suture at the junction of the segments is usually either indistinct or wanting. The segments are quite entire.

M. de Brébisson has constituted this genus for some plants which he separated from *Closterium* on account of their straight fronds, which have the opposite margins similar. It differs from *Tetmemorus* by the absence of a terminal notch, and from *Cosmarium* by its more elongated frond and its less marked constriction at the junction of the segments; yet the cylindrical species of *Cosmarium* closely approach to this genus, and *Cosmarium turgidum* and *Cosmarium attenuatum* are at least as much elongated as *Penium truncatum*.

Although the species in *Penium* are few, they present considerable differences in their characters. Some are smooth, and, like *Closterium*, have the endochrome arranged in longitudinal fillets, and at each extremity of it a globule containing moving granules. Other species have a granulated surface, and there are two smooth and truncate which some botanists have referred to the Palmelleæ.

Penium Digitus, on account of its size, is represented as magnified only 200 diameters.

* *Empty frond either striate or granulate, generally reddish.*

1. *P. margaritaceum* (Ehr.); frond cylindrical with rotundato-truncate ends; pearly granules in longitudinal series.

> α. Frond fusiform, constricted at the middle; granules distinct, giving a denticulate appearance to the margin.
>
> β. Frond linear, scarcely constricted at the middle; granules giving a denticulate appearance to the margin, somewhat scattered near the suture.
>
> γ. *punctatum*; frond linear, neither constricted at the suture nor denticulate at the margin; granules appearing like puncta.

Closterium margaritaceum, Ehrenberg, *Infus.* p. 95. t. 6. f. 13 (1838). Meneghini, *Synop. Desmid.* p. 236. Jenner, *Fl. of Tunbridge Wells*, p. 196. Kützing, *Phy. Germ.* p. 132. Hassall, *Brit. Freshwater Algæ*, p. 376. *Penium margaritaceum*, Brébisson, *in lit.* (1846).

α. Ashdown Forest, Sussex, *Mr. Jenner.* Penzance, *J. R.*

β. Near Bristol, *Mr. Thwaites.* Warbleton, Sussex, *Mr. Jenner.*

γ. Dolgelley, *J. R.* Ashdown Forest, and Henfield, Sussex; and Reigate, Surrey, *Mr. Jenner.* Near Aberdeen, *Dr. Dickie* and *Mr. P. Grant.* Ambleside, *Mr. Sidebotham.*

Berlin, *Ehrenberg.* Falaise, *Brébisson.*

Frond minute, six to twelve times longer than broad, rough with pearly granules which are arranged in longitudinal rows. The forms which I have here considered varieties differ so much, that it is probable they will prove to be distinct species. It will be seen that the variety β. is intermediate between the extreme states, and since neither of our British forms agrees with Ehrenberg's figure and description, I shall describe each in detail.

α. Frond stout, rounded at the end, distinctly constricted at the middle; fillets distinct, interrupted and, as well as the endochrone, divided into four equal portions by three pale transverse bands. Near each extremity there is an obscure globule; segments somewhat fusiform. The empty frond is nearly colourless, and the pearly granules are distinct and give a denticulate appearance to the margin. Besides the middle transverse suture, there are two others corresponding with the interruptions of the endochrome. I have received this form from M. de Brébisson as his *Penium margaritaceum.* It differs from Ehrenberg's figure of *Closterium margaritaceum* in being broader, more constricted at the middle, and having more ventricose segments. The arrangement of the endochrome and the position of the globules also differ.

The variety β. agrees in colour with the preceding, but it is shorter, and its sides are parallel. The granules are distinct, but less regular. I have only seen it after the endochrome had collapsed.

The variety γ. is elongated, slender, linear, truncate at the ends, and scarcely constricted at the middle. The endochrome is divided into four portions by transverse pale bands; but there are no fillets or globules with moving particles. The empty frond is reddish, especially at its extremities, the last-formed segment being generally paler than the other. The pearly granules are very minute, and resemble puncta; hence the margin of the frond is entire. The puncta or granules are arranged in longitudinal lines, but those in the same line are so close together that they are nearly, if not quite, confluent; in fact, unless a very high power of the microscope be used, the frond appears simply striate. Besides the central suture, several other transverse lines divide the empty frond into compartments. This form agrees best with Ehrenberg's figure of *Closterium margaritaceum*, but it is not denticulate at the margin.

Sporangia of γ. have been gathered by myself near Dolgelley, and those of β. by Mr. Thwaites near Bristol. Both are orbicular, and situate between the deciduous fronds.

Length of frond in α. $\frac{1}{156}$ of an inch; breadth $\frac{1}{961}$; breadth at constriction $\frac{1}{1234}$; length of frond in γ. $\frac{1}{109}$; breadth $\frac{1}{1515}$; diameter of sporangium $\frac{1}{543}$.

Tab. XXV. fig. 1. *a*. frond of α. with endochrome; *b, c*. empty fronds; *d*. empty frond of γ. as seen under a low power; *e, f*. fronds with endochrome; *g, h*. sporangia.

Tab. XXXIII. fig. 3. sporangium of β.

2. **P. Cylindrus** (Ehr.); frond cylindrical with rotundato-truncate ends, rough with scattered pearly granules.

Closterium? Cylindrus, Ehrenberg, *Infus*. p. 95. t. 6. f. 6 (1838). Meneghini, *Synop. Desmid*. p. 236. Kützing, *Phy. Germ*. p. 132.
Penium Cylindrus, Brébisson, *in lit*. (1846).

Dolgelley, *J. R.* Ambleside, *Mr. Sidebotham*. Henfield; Ashdown Forest, and Piltdown Common near Uckfield, Sussex, *Mr. Jenner*.

Berlin, *Ehrenberg*. Falaise, *Brébisson*.

Frond very minute, smaller than that of *Penium margaritaceum*, two or three times longer than broad, cylindrical with truncate ends, and rough with minute scattered granules or apiculi. As the covering is usually of a deep rusty red, it obscures the view of the endochrome. I have not observed any terminal globules.

Penium Cylindrus resembles *P. margaritaceum* in form, but is much smaller, and its pearly granules are scattered and not arranged in lines. The red colour of the empty frond is darker than that of any other plant in the family.

I have little hesitation in referring the British specimens to Ehrenberg's *Closterium Cylindrus*, although in their scattered granules they differ from his and Meneghini's descriptions.

Length of frond $\frac{1}{492}$ of an inch; breadth $\frac{1}{1760}$.

Tab. XXV. fig. 2. *a*. frond with endochrome; *b*. dividing frond; *c*. empty frond.

** *Frond smooth, colourless.*

† *Sporangium situated between the deciduous cells.*

3. **P. Digitus** (Ehr.); frond smooth, elliptic-oblong with rounded ends; fillets obscure, undulated, interrupted only by a transverse band at the suture.

Closterium Digitus, Ehr. *Abh. der Berl. Ak.* (1831), p. 68 ; *Infus.* p. 94.
t. 6. f. 3 ? Meneghini, *Synop. Desmid. in Linnæa* 1840, p. 236. Bailey,
Amer. Bacil. in Amer. Journ. of Science and Arts, v. 41. p. 302. t. 1. f. 33.
Jenner, *Fl. of Tunbridge Wells*, p. 196. Hassall, *Brit. Alg.* p. 376.
Kützing, *Phy. Germ.* p. 132.
Closterium lamellosum, Brébisson, *Alg. Fal.* p. 59. t. 8 (1835).
Pleurosicyos myriopodus, Corda, *Alm. de Carlsb.* 1835, p. 125. t. 5. f. 68.
Penium Digitus, Brébisson, *in lit.* (1847).
Polysolenia Closterium, Ehr. *MS.* Bailey, *in lit.*

Very common. Wales and Penzance, *J. R.* Sussex, Surrey, Kent and Hants, *Mr. Jenner.* Kerry, *Mr. Andrews.* Near Bristol, *Mr. Thwaites.* Aberdeenshire, *Dr. Dickie* and *Mr. P. Grant.* Banffshire, *Mr. P. Grant.* Ambleside, *Mr. Sidebotham.*

Germany, *Ehrenberg, Kützing, Corda.* Falaise, *Brébisson.* Maine to Virginia, *Bailey.*

Frond smooth, stout, three to five times longer than broad, elliptic or elliptic-oblong; not unfrequently however the frond is dilated at one end and becomes ovato-oblong; the ends are broadly rounded. The endochrome is yellow-green with a pale transverse band at the middle; there are no regular fillets, but longitudinal undulations are more or less perceptible. Terminal globules wanting, or very indistinct, in British specimens. The end view is circular, and the endochrome radiate. The empty frond is colourless, flexible, and without markings.

Ehrenberg's description and figure of his *Closterium Digitus* is unlike the plant here described, and Brébisson formerly supposed it was intended for the next species; but the whole outline and the description of the rounded ends are certainly more suitable to this. I have never seen any terminal vesicles, yet they are depicted by Ehrenberg and also by Professor Bailey in a figure in his 'American Bacillaria.'

In the recent state there can be no danger of confounding *Penium Digitus* with *P. interruptum*; and even in the dried state, although the conspicuous differences dependent on the disposition of the endochrome are no longer to be found, the form of the frond appears to me a sufficient distinction.

Length of frond $\frac{1}{81}$ of an inch; greatest breadth $\frac{1}{299}$.

Tab. XXV. fig. 3. *a, b.* fronds with endochrome; *c.* empty frond.

4. **P. interruptum** (Bréb.); frond smooth, cylindrical; extremities conical with rounded apex; fillets strongly marked, interrupted by three transverse bands.

Penium interruptum, Brébisson, *in lit.* (1846).

Dolgelley and Penzance, *J. R.* Ashdown Forest and Chiltington Common, Sussex, *Mr. Jenner.* Ambleside, Westmoreland, *Mr. Sidebotham.*

Falaise, *Brébisson.*

Frond stout, four to six times longer than broad, quite straight, cylindrical, so that the sides are parallel, except at the extremities, which are rather suddenly conical and rounded at the apex. Endochrome dark green; fillets straight, three or four strongly marked and a fainter one near each margin.

Besides the central transverse band present in all the Desmidieæ, *Penium interruptum* has two others; the fillets and endochrome are consequently divided into four portions. The ends are colourless, and each is partly occupied by a large globule, which contains moving granules like those in *Closterium*. The empty frond is colourless without markings, and there is no transverse line at the middle.

In a recent state this is a remarkable plant. The fillets are more strongly marked than in any other species I have met with, and the checkered appearance produced by the transverse bands essentially distinguish it. It must however be observed that this character is lost in the dried state, when its straight, cylindrical fronds have some similarity to those of *Penium Digitus*; but they are longer in proportion, their sides more parallel, and their extremities more conical.

Length of frond $\frac{1}{116}$ of an inch; breadth $\frac{1}{571}$.

Tab. XXV. fig. 4. *a*. frond magnified 400 times; *b*. frond magnified 200 times.

5. **P. closterioides** ——; frond smooth, fusiform or lanceolate; longitudinal fillets and terminal globules distinct.

Penzance and Dolgelley, *J. R.* Ashdown Forest, and Piltdown, Sussex; Reigate, Surrey; and near Southampton, *Mr. Jenner*.

Frond six to ten times longer than broad, straight, lanceolate; end obtuse; fillets and terminal globules distinct; vesicles in a single longitudinal series. Empty frond colourless and without markings.

Penium closterioides in its appearance and in the arrangement of its endochrome agrees better with *Closterium* than with *Penium*, and strongly resembles the side view of some species of the former. The frond however is always straight, and the opposite margins are symmetrical.

Length of frond $\frac{1}{92}$ of an inch; breadth $\frac{1}{590}$.

Tab. XXXIV. fig. 4. *a*. frond with endochrome; *b*. empty frond.

6. **P. truncatum** (Bréb.); frond smooth, cylindrical with truncate ends; sporangia orbicular, easily detached from the conjugated fronds.

Cylindrocystis truncata, Brébisson, *in lit. cum icone* (1847).

Dolgelley, *J. R.*

Falaise, *Brébisson*.

Frond very minute, two or three times longer than broad, smooth, cylindrical with truncate ends, sometimes slightly constricted at the middle. The endochrome is similar to that of *Penium Brebissonii*, but the transverse pale band at the middle is usually more distinct.

The empty frond is colourless, and its suture imperceptible. I believe that it is minutely dotted, but the dots are too indistinct to assist in defining the species.

I have gathered the sporangia rather abundantly in a pool near Dolgelley;

they are orbicular and situated between the conjugating fronds, from which, after their formation, they soon become detached.

Penium truncatum closely resembles a small specimen of *P. Brebissonii*, but it is much smaller and its ends are truncate. In form it agrees with *P. Cylindrus*, but in that species the empty frond is coloured and distinctly granulate.

Length of frond from $\frac{1}{969}$ to $\frac{1}{555}$ of an inch; breadth from $\frac{1}{2212}$ to $\frac{1}{2100}$; diameter of sporangium $\frac{1}{950}$.

Tab. XXV. fig. 5. *a, b, c.* mature fronds; *d.* dividing frond; *e.* conjugated fronds; *f, g.* sporangia.

7. **P. Jenneri** ——; frond smooth, cylindrical with rounded ends; sporangium orbicular, situated between the conjugating fronds, which are deciduous.

Bogs at Fisher's Castle, Tunbridge Wells, *Mr. Jenner.*

Frond variable in length, from two to four times longer than broad, oblong or cylindrical with rounded ends.

I know no character by which to distinguish the usual state of *Penium Jenneri* from *P. Brebissonii*. They agree in size and form; in both the arrangement of the endochrome is similar, and the transverse pale band at the middle is often indistinct; the empty frond also is smooth and colourless, and no suture can be detected. Still the different forms of their sporangia compel us to consider them as distinct species.

In the present plant the conjugating fronds do not open and gape at the suture, as is usual in the Desmidieæ, but couple by small and distinct cylindrical tubes, like many in the Conjugatæ. The sporangium is orbicular, and situated between the fronds, within the dilated connecting tube.

The frond in *Penium Jenneri* is more rounded at its ends and larger than those of *P. truncatum*; in the latter species, moreover, the conjugating cells gape at the suture; in both they are but slightly attached to the sporangium.

Length of frond from $\frac{1}{810}$ to $\frac{1}{433}$ of an inch; breadth from $\frac{1}{1724}$ to $\frac{1}{1666}$; diameter of sporangium $\frac{1}{1000}$.

Tab. XXXIII. fig. 2. *a, b, c, e.* fronds with endochrome; *d.* empty frond; *f.* dividing frond; *g.* conjugated fronds; *h, i.* mature sporangia.

†† *Sporangium permanently attached to the conjugated cells, at first quadrate, then orbicular* (Cylindrocystis).

8. **P. Brebissonii** (Menegh.); frond smooth, cylindrical with rounded ends, transverse central band inconspicuous; sporangium at first quadrate but finally orbicular; conjugating fronds persistent.

Palmella cylindrospora, Brébisson, *Alg. Fal.* p. 64 (1835).
Cylindrocystis Brebissonii, Meneghini, *Cenni sulla organografia e fisiol. delle Alg.* pp. 5, 26 (1838); *Monog. Nost.* t. 12. f. 3. Hassall, *Brit. Freshwater Alg.* p. 361. t. 92. f. 17.
Cymbella? lætevirens, Harvey, *Man. of Brit. Alg.* p. 216 (1841).
Closterium Cylindrocystis, Kützing, *Phy. Germ.* p. 132 (1845).

Very common, especially where water lodges in wet weather. Cornwall and Wales, *J. R.* Ayrshire, *Rev. D. Landsborough.* Aberdeenshire and Banffshire, *Dr. Dickie* and *Mr. P. Grant.* Northamptonshire, *Rev. M. J. Berkeley.* Sussex, Kent, Surrey, and near Southampton, *Mr. Jenner.* Ireland, *Mr. Andrews.* Essex, *Mr. Hassall.* Near Bristol, *Mr. Thwaites* and *Mr. Broome.* Westmoreland, *Mr. Sidebotham.* Rochdale, *Mr. Coates.*

Falaise, *Brébisson.* Germany, *Kützing.* Princeton, New Jersey, with sporangia, *Bailey.*

Frond minute, often congregated into a mucous stratum on wet mud, in the same manner as *Tetmemorus granulatus,* its frequent companion, which at first sight it resembles. The fronds vary much in size, from two to six times longer than broad, and are oblong or cylindrical with rounded ends. The transverse pale central band, which in the Desmidieæ indicates the suture or junction of the segments, is less conspicuous here than in any other plant of the family. The endochrome being usually denser at the middle of each segment produces a resemblance to a joint of a species of *Tyndaridea.* The end view is circular with a central nucleus. The empty frond is smooth and colourless; no suture can be detected in it. The sporangia have been gathered by me at Dolgelley; by Mr. Jenner on Piltdown Common and near Tunbridge Wells, Sussex, and by Mr. Thwaites near Bristol. The process of conjugation in this species differs from that in the rest of this genus; for, as in *Hyalotheca dissiliens,* the conjugating cells enter into the formation of the containing cell and are permanently attached to the sporangium, instead of being at length detached, as commonly happens in the Desmidieæ. The sporangium is at first cruciform, then quadrate, and finally orbicular.

Penium Brebissonii has been placed with the Palmelleæ by Meneghini and Brébisson, because the division into two segments is but obscurely indicated, and its fronds often form a mucous stratum. I cannot however concur with them, and I consider that it has been correctly removed by Kützing to this family. *Tetmemorus granulatus, Cosmarium curtum,* and other undoubted Desmidieæ, form similar strata, are often mingled with it, and by the naked eye cannot be distinguished from it. In its formation of sporangia this plant also agrees with the undoubted members of the family.

That this species has been referred to the Palmelleæ by such distinguished algologists proves indeed how closely the families are allied. M. de Brébisson, in a recent letter, admits that it may belong to the Desmidieæ, but considers that the difference in the formation of its sporangia is a sufficient reason for keeping the genus. Admitting that in strictness his view is correct, I nevertheless place this species in *Penium,* because the conjugated state, occurring so rarely, would seldom enable the student to distinguish between *Penium* and *Cylindrocystis.*

Length of frond from $\frac{1}{637}$ to $\frac{1}{404}$ of an inch; breadth from $\frac{1}{1666}$ to $\frac{1}{578}$; length of side of quadrate sporangium from $\frac{1}{777}$ to $\frac{1}{694}$; diameter of mature sporangium $\frac{1}{470}$.

Tab. XXV. fig. 6. *a, b.* fronds with endochrome; *c, d.* empty fronds; *e, f, g.* imperfect sporangia; *h, i.* mature sporangia.

15. DOCIDIUM, *Bréb.*

Frond simple, straight, much elongated, linear, constricted at the middle and truncate at the ends; *segments* usually inflated at the base.

Fronds simple, straight, much elongated, sometimes even twenty times longer than broad, constricted at the middle, where the suture is generally very distinct in both living and empty fronds. In almost every species the segments are inflated at their base; the ends are truncate.

Professor Bailey has been very successful in adding new and curious species to this genus, some of them differing greatly from any previously known.

Docidium, like *Closterium*, has terminal globules containing moving granules, and its vesicles are either scattered or arranged in a single longitudinal row.

It differs from *Closterium* in its straight fronds and constricted middle; and from *Penium* it may be known by having a distant constriction at the middle and more elongated fronds.

1. *D. nodulosum* (Bréb.); frond stout; segments four to six times longer than broad, constricted at regular intervals so as to produce undulated margins; suture projecting on each side.

Closterium Trabecula, Bailey, *Amer. Bacil. in Amer. Journ. of Science and Arts*, v. 41. p. 302. t. 1. f. 32? (1841).
Docidium nodulosum, Brébisson, *in lit. cum icone* (1846).

Dolgelley, *J. R.* Durdham Down near Bristol, *Mr. Thwaites.* Aberdeenshire, *Dr. Dickie* and *Mr. P. Grant.* Ambleside, Westmoreland, *Mr. Sidebotham.* Henfield, Sussex, and near Southampton, *Mr. Jenner.*

Falaise, *Brébisson.* Maine to Virginia, *Bailey.*

Frond large, distinctly visible to the naked eye, beam-like, constricted at the middle, where the thickened suture projects on each side like an apiculus. Segments about four times as long as broad, scarcely attenuated, regularly inflated at intervals, whence the margins appear undulated; the basal inflation is the most prominent, the others less so as they are situated nearer the end; often only two or three nearest the base are distinct, but occasionally as many as eight undulations may be counted at the margin.

Docidium nodulosum differs from *D. constrictum*, Bailey, and *D. nodosum*, Bailey, in its more numerous but slighter constrictions. In both these species the inflations are always four, conspicuous, and more distant. In *D. verrucosum*, Bailey, the segments are longer in proportion to their breadth; the undulations, which are more numerous and equal, depend on the presence of whorls of small tubercles. In *D. coronatum*, Bréb., the apex being encircled by a series of small tubercles, the end view is crenate.

Length of frond $\frac{1}{50}$ of an inch; breadth $\frac{1}{428}$; breadth at inflation $\frac{1}{393}$; breadth at end $\frac{1}{625}$.

Tab. XXVI. fig. 1. *a*. frond with endochrome; *b*. empty frond.

2. *D. truncatum* (Bréb.); frond stout; segments three or four times longer than broad, with a single inflation at the base, tapering to the truncate entire ends; suture projecting on each side.

Closterium truncatum, Bréb. Meneghini, *Synop. Desmid. in Linnæa* 1840, p. 235.
Docidium truncatum, Brébisson, *in lit. cum icone* (1846).

Carnarvon and Penzance, *J. R.* Hastings, Sussex, and Brookland near Reigate, Surrey, *Mr. Jenner*.

Falaise, *Brébisson*. Germany, *Kützing*.

Frond cylindrical, stout, distinctly constricted at the middle, the suture forming a thickened rim, which projects on each side like a small point. Segments three or four times longer than broad, slightly inflated at the base and tapering towards the end, which is truncate and entire.

Docidium truncatum seems to me to approach closely to *D. nodulosum*, with which it agrees in size. Its distinctive marks are the more attenuated extremities and the solitary inflation at the base of each segment, so that its margins are not undulated. Mr. Jenner considers them as undoubtedly distinct, but to which dried specimens belong I am occasionally unable to decide; for in drying, fronds of *D. nodulosum* frequently become attenuated at the ends, and sometimes only the basal inflation can be seen. Its larger size, shortness of the segments in comparison with their breadth, and the distinct projection of the junction-rim on each side, sufficiently distinguish *D. truncatum* from *D. Ehrenbergii*.

Length of frond from $\frac{1}{81}$ to $\frac{1}{72}$ of an inch; greatest breadth from $\frac{1}{526}$ to $\frac{1}{439}$; breadth at constriction $\frac{1}{591}$; breadth at end from $\frac{1}{1168}$ to $\frac{1}{940}$.

Tab. XXVI. fig. 2. *a*. frond with endochrome; *b*. empty frond.

3. *D. clavatum* (Kütz.); segments many times longer than broad, with a single inflation at the base; end clavate, but truncate at the apex.

Closterium truncatum, Kützing, *Phy. Germ.* p. 133 (1845), nec Bréb.
Docidium clavatum, Kützing, *in lit. cum icone* (1846). Brébisson, *in lit. cum icone et specimine*.

Midhurst, Sussex, *Mr. Jenner*.

Germany, *Kützing*. Falaise, *Brébisson*.

Frond longer than those of the first two species, but more slender, constricted at the middle; the suture strongly marked, but not projecting at the margin.

Segments eight to twelve times longer than broad, gradually thickened near the ends in a clavate manner, the end itself being slightly attenuated and truncate; there is a single inflation at the base. Vesicles small, few, and arranged in a single longitudinal line. The empty frond is minutely punctate, but the highest powers of the microscope are required to detect the puncta.

Brébisson's specimens are more clavate than British ones, or than Kützing's plant appears in a drawing received from him.

Length of frond $\frac{1}{65}$ of an inch; breadth $\frac{1}{813}$; breadth at inflation $\frac{1}{909}$; breadth at end $\frac{1}{1333}$.

Tab. XXVI. fig. 3. *a.* frond with endochrome; *b.* empty frond from French specimen.

4. *D. Ehrenbergii* ———; segments many times longer than broad, having two or three slight inflations at the base and truncate ends bordered by minute tubercles.

β. *granulatum*; frond rough with minute granules.

Closterium Trabecula, Ehr. *Abh. d. Berl. Ak.* (1830), p. 62; *Infus.* p. 93. t. 6. f. 2. Meneghini, *Synop. Desmid.* p. 235. Kützing, *Phycologia Germanica*, p. 133? Jenner, *Flora of Tunbridge Wells*, p. 196.

Dolgelley, Carnarvon, Penzance, &c., *J. R.* Several stations in Sussex; and near Southampton, *Mr. Jenner.* Aberdeenshire, *Dr. Dickie* and *Mr. P. Grant.* Ambleside, *Mr. Sidebotham.*

β. Henfield, Sussex, *Mr. Jenner.* Penzance, *J. R.*

Germany, *Ehrenberg, Kützing.* Falaise, *Brébisson.*

Frond as long as those of *Docidium nodulosum* and *D. truncatum*, but far more slender, constricted at the middle; the suture strongly marked, but having no projection at the margins. Segments eight to twelve times longer than broad, having about two slight inflations at the base, but otherwise linear. The end is truncate, and three to five minute tubercles, which may be detected at its margin, give the end view a crenate appearance. The empty frond is punctate, but in general the puncta are not readily discerned. Mr. Jenner finds in Sussex a variety the frond of which is distinctly granulate.

I have gathered the sporangia at Penzance for two successive years; they are large, suborbicular or elliptic, but often slightly angular, and are situated between the conjugating fronds, from which they easily separate.

Docidium Ehrenbergii in size equals *D. clavatum*, but it is not clavate, and its ends are bordered by minute tubercles. In the latter character it corresponds with *D. coronatum*, Bréb.; but it is much smaller and the suture does not project at the sides.

Length of frond from $\frac{1}{71}$ to $\frac{1}{59}$ of an inch; breadth from $\frac{1}{1111}$ to $\frac{1}{961}$; breadth at inflation from $\frac{1}{909}$ to $\frac{1}{881}$; breadth at end $\frac{1}{1184}$; diameter of sporangium $\frac{1}{285}$.

Tab. XXVI. fig. 4. *a.* frond with endochrome; *b.* dividing frond; *c.* empty frond; *d.* conjugating fronds; *e, f.* sporangia.

Tab. XXXIII. fig. 4. empty frond of variety β.

5. *D. Baculum* (Bréb.); segments very slender, having a solitary and conspicuous inflation at the base, otherwise linear; vesicles in a single series.

Closterium Baculum, Brébisson, *Alg. Fal.* p. 59. t. 8 (1835).
Closterium trabeculoides, Corda, *Observ. Microscop. sur les Animal. de Carlsbad*, t. 6. f. 44 (1840).
Closterium Sceptrum, Kützing, *Phycologia Germanica*, p. 133 (1845).
Docidium Baculum, Brébisson, *in lit. cum icone* (1846).

Common. Carvarvon; Dolgelley; near Lampeter; Penzance, &c., *J. R.* Ashdown Forest; near Storrington, &c., Sussex, *Mr. Jenner.* Near Bristol, *Mr. Thwaites* and *Mr. Broome.* Aberdeenshire, *Dr. Dickie* and *Mr. P. Grant.* Ambleside, *Mr. Sidebotham.* Rochdale, *Mr. Coates.*

Falaise, *Brébisson.* Carlsbad and Prague, *Corda.* United States, *Bailey.*

Frond very slender, many times longer than broad, distinctly constricted at the middle. Suture strongly marked, but having no projections at the margins. Segments with a solitary inflation at the base, otherwise linear, the ends truncate and without tubercles.

Docidium Baculum may be readily recognized, as it is more slender than any of the preceding species, and its inflations are more developed.

Length of frond $\frac{1}{111}$ of an inch; greatest breadth $\frac{1}{1937}$; breadth at inflation $\frac{1}{1938}$; breadth at end $\frac{1}{2463}$.

Tab. XXXIII. fig. 5. *a, b.* fronds with endochrome; *c.* empty frond magnified 400 times.

6. *D. minutum* ——; frond slender, slightly constricted at the middle; segments linear, not inflated; vesicles in a single series.

Dolgelley, *J. R.* New Forest, Hants; and Parham Old Park near Storrington, Sussex, *Mr. Jenner.*

Frond minute, smaller than those of any of the foregoing species, linear, the constriction often scarcely perceptible; segments four to six times longer than broad, the ends truncate.

The smaller size and uninflated state of *Docidium minutum* distinguish it from *D. Baculum.*

Length of frond $\frac{1}{212}$ of an inch; greatest breadth $\frac{1}{1582}$.

Tab. XXVI. fig. 5. *a.* frond with endochrome; *b.* empty frond.

7. *D.? asperum* (Bréb.); frond slender, cylindrical, rough with minute scattered granules, neither constricted nor inflated at the middle; ends truncate and dilated.

Docidium asperum, Brébisson, *in lit. cum icone* (1846).

Kerris near Penzance, *J. R.*

Falaise, *Brébisson.*

Frond slender, twelve or more times longer than broad, rough with scattered minute points which are best seen at the margins. Endochrome irregular. I have seen neither vesicles nor terminal globules.

I am doubtful whether this plant be a true *Docidium*, for in all the other species of this genus there is, at the middle, an evident constriction, and a conspicuous suture where the segments eventually separate; on the other hand, this plant has neither constriction nor suture.

Specimens of *Docidium asperum* sent me by M. de Brébisson are attenuated towards the extremities, whilst the apex itself is dilated. British specimens are not constricted beneath the dilated ends. Brébisson finds moving granules at the extremities as in the other species, but I have not noticed them.

This plant is not unlike the separated joints of a species of *Mougeotia*, but the dilated ends and roughness produced by minute granules distinguish it.

Length of frond from $\frac{1}{97}$ to $\frac{1}{64}$ of an inch; breadth $\frac{1}{2356}$; breadth at end $\frac{1}{2272}$.

Tab. XXVI. fig. 6. *a.* frond with endochrome; *b.* empty frond; *c.* empty frond of French specimen.

16. CLOSTERIUM, *Nitzsch*.

Frond simple, elongated, attenuated, lunately curved or arcuate, entire; the junction of the *segments* marked by a pale transverse band.

In *Closterium* the frond is simple, much elongated, fusiform, rarely cylindrical or prismatic, more or less attenuated at the extremities. It is semilunate, crescent-shaped, or, in a few instances, curved only at the ends, but in the usual position the opposite margins are always more or less dissimilar, the upper one being convex, and the lower, inclusive of the ends, straight, or more commonly concave; the lateral view, which differs from the front one, is straight and has both margins similar. As happens in the allied genera, the frond is composed of two segments which finally separate and allow the escape of the endochrome; the suture or junction is marked by a transverse hyaline or pale band, accompanied in some species by one or more transverse striæ, best seen in the empty frond; but there is no constriction. The segments are entire, without spines, processes, or inflated protuberances.

The endochrome is green with darker longitudinal bands or fillets, the number of which varies in different species. Meneghini considers them of too much importance to be omitted in the specific definition*. They may occasionally be useful in discriminating nearly allied forms,

---

* " Interanea initio uniforma serius in tænias distribuuntur, quarum characteres constantes videntur et in specifica definitione omitti nequeunt."— *Menegh. Synop. Desmid.* p. 230.

but as they are frequently indistinct, or from various causes it may be difficult to count them with certainty, I am unwilling to introduce them into the specific character except in the absence of more permanent marks of distinction.

The diaphanous vesicles, which are conspicuous in most of the species, are either scattered or arranged, with more or less regularity, in a single longitudinal series.

At each extremity of the endochrome, even in its earliest state, there is a large hyaline or straw-coloured globule, which contains minute granules in constant motion. This globule disappears in the dried specimen.

A distinct circulation has been noticed in several species. It is said to occur only in specimens obtained from water, and not in those taken from moist ground.

The empty frond is striated in some species and smooth in others, but no instance of its being granulate is known. The striated species are by Ehrenberg placed together in a subgenus (*Toxotium*), but it is sometimes very difficult to detect the striæ, which I have also seen distinctly in *Closterium attenuatum*, and other species usually considered smooth.

The colour and firmness of the covering differ in different species. Some are quite colourless, flexible, collapsing when dried, and in general allowing the endochrome to escape by a merely partial separation of the segments. These species are never striated. In other species the fronds are more or less straw-coloured or even reddish, probably from the presence of a small portion of silicate of iron. The deeper the colour the firmer is the frond. The segments separate entirely from each other and retain their shape when empty or dried; and some of the striated species, even when submitted to the action of nitric acid or fire, retain their form and markings. In the coloured species the extremities are generally darker than the rest of the frond.

Even in the firmest species the frond becomes flattened in drying, its breadth at the centre increases, and the ends appear more attenuated than in their living state. This fact should be attended to when describing or drawing a dried specimen.

In *Closterium*, of which several species have been noticed in a conjugated state, the process appears to be nearly the same as in the Conjugatæ. Two fronds unite by means of projections arising at the junction of the two segments, and then the newly-formed portion continues to enlarge until the original segments are separated by a cell of an irregular four-sided figure. The contents of the fronds, being collected in this cell, become a dense seed-like mass, which is sometimes globular resembling the sporangium in *Mougeotia*, and sometimes square like that in *Staurospermum*. The newly-formed cell is thinner and generally paler than the segments of the frond; in some species it looks like a prolongation of the segments, and in others these are so loosely attached that their connection is scarcely perceptible.

Two species having cruciform or quadrate sporangia permanently attached to the empty segments of the conjugated fronds are separated by Kützing to form a new genus, *Stauroceras*. To these others have been added by Brébisson. Whilst fully admitting the soundness of the principle on which this separation has been made, I regret that I am unable to adopt it here, not only because I am uncertain to which genus some of the species should be referred, but because I have decided to employ in this monograph only such generic characters as can be determined in the ordinary state of the plant, and will not oblige the observer to depend on conditions which he may not be able to discover.

According to Meneghini the coupling of the fronds takes place from the convex margin. This is generally, but not invariably, the case, for I have observed that they are connected sometimes on the convex and sometimes on the concave margin; in some instances, indeed, I have seen the convex margin of one frond connected with the concave margin of another. Mr. Jenner has noticed the same diversity, but it appears to him to characterize different species.

As there is no constriction in *Closterium*, although its frond divides like that of other Desmidieæ, the process of division is less evident. It is best seen in the striated species, in which the central suture is most distinct. The transverse line becomes double, and by the intermediate growth the frond at length consists of three portions. As the newly-formed central one continues to elongate, another transverse line becomes visible at its middle, where a complete separation at length takes place. At first however the new segment is often merely a rounded protuberance, and the frond is consequently unequal; and when it is perfected, if the covering is a coloured one, the newer segment can still be distinguished by its greater paleness.

The striated species, besides the central suture, frequently have other transverse lines that divide the segments themselves into two or more portions. Brébisson considers that their number is constant in the same species, and that they may be used in framing a specific character; but my experience leads me to agree with Meneghini, that it is generally unsafe to place much reliance upon them: nevertheless it must be admitted that in many species no other suture than the central one is ever present, except during the process of division.

Closterium may be distinguished from all the other genera of the Desmidieæ by its elongated, curved, entire, and attenuated fronds.

I have divided *Closterium* into sections answering to the genera of Kützing and Brébisson, according to a list of the species sent me by the latter; but I must notice that one or two species will probably be found to agree just as well with some other section as with that in which it is placed.

In the first section the fronds are commonly either semilunate or crescent-shaped, and have the extremities blunt or but moderately elongated; the terminal globules are situated so near the ends that the endochrome nearly fills the cell; the extremities appear to have a slight notch or depression (the opening of Ehrenberg); the sporangia are orbicular, and are situated between the conjugating fronds and but slightly connected with them.

The second section contains but one species, which by Brébisson is still retained in *Closterium*; the form however of its frond agrees with that of Kützing's *Stauroceras*, whilst its sporangium is unlike that of any other species. The fronds are inflated at the middle and elongated at each end into a short curved beak, and there is no notch apparent at the ends. The sporangium, which is placed between the conjugating cells, is geminate, consisting of two nearly orbicular portions, flattened at their junction, where they often separate.

The plants in the third section are in general curved only at their extremities, which taper considerably, and frequently form a slender beak. The margins of the frond, exclusive of the curved extremities, are in most cases nearly similar; the moving granules are more or less distant from the extremities, which have no distinct terminal notch; and the segments are permanently attached to the sporangium, which is cruciform or quadrate.

* *Sporangium orbicular, situated between the conjugating fronds and but slightly connected with them; fronds never rostrate.* (Closterium.)

† *Frond semilunate or semilanceolate, tapering from the middle, the lower margin straight (or nearly so) and inclined upwards at the end.*

1. *C. Lunula* (Müller); frond smooth, semilunate; lower margin nearly straight, inclined upwards at the rounded ends; vesicles numerous, scattered.

Vibrio Lunula, Müller, *Naturforsch.* p. 142 (1784); *Animal. Infus.* p. 55. t. 7. f. 13 and 15.
Mulleria Lunula, Leclerc, *Mém. du Mus. I.* (1802). Schrank, *Faun. Boica*, 3. p. 47.
Bacillaria Lunula, Schrank, *Acta N. Cur.* 11. p. 533 (1823).
Lunulina vulgaris, Bory, *Encyclop. Méth. Hist. Nat. des Zooph.* t. 2 (1824). Turpin, *Dict. d'Hist. Nat.* t. 5.
Closterium Lunula, Ehr. et Hemprich, *Symb. Phys.* 1828, t. 2. Ehr. *Abh. der Berl. Ak.* 1830; *Infus.* p. 90. t. 5. f. 15. Kützing, *Alg. aq. dulc.* No. 22; *Synopsis Diatom. in Linnæa* 1833, p. 596. Corda, *Alm. de Carlsbad*, 1835, p. 190. t. 5. f. 56 and 57. Jenner, *Fl. of Tunbridge Wells*, p. 196.

Common. Carnarvon; Dolgelley; Penzance, &c., *J. R.* Several stations in Sussex, Kent, and Surrey, *Mr. Jenner.* Kerry, *Mr. Andrews.* Aberdeenshire, *Dr. Dickie* and *Mr. P. Grant.* Near Congleton, Cheshire; Ambleside, and near Manchester, *Mr. Sidebotham.*

Germany, *Kützing, Ehrenberg, Corda.* Falaise, *Brébisson.* Common in New York and New England; Mexico, *Bailey.*

Frond bright green, stout, distinctly visible to the naked eye, semilunate, five or six times longer than broad; extremities conic with rounded and very obtuse ends. The upper margin is very convex, the lower straight, except at the extremities which incline upwards, so that the segments rapidly taper from their junction. Endochrome grass-green; fillets several, three more distinct than the rest; vesicles numerous, small and scattered.

The empty frond is colourless and without markings, and its suture is indistinct or wanting. It shrinks in drying, and is destroyed by burning.

Closterium Lunula differs from *C. Ehrenbergii* and *C. moniliferum* in having no inflation at the middle of the lower margin; and although, in drying, it often acquires a prominent centre, its upward inclination at the extremities is a permanent mark of distinction. It may be known from *C. acerosum* by its stouter appearance and scattered vesicles.

I have never seen a specimen in which the length exceeded the breadth so much as in Ehrenberg's figure.

Meneghini refers *C. Lunula* of Kützing's 'Alg. Aq. Dul.' to *Closterium moniliferum*; but two specimens which I have had an opportunity of examining, one in a copy of that publication, and another given me by Kützing himself, belong to this species. I have therefore followed Ehrenberg in considering the present as the plant of Nitzsch, whilst the figure in Kützing's 'Synopsis Diatomearum' must be referred to *C. moniliferum.*

Length of frond $\frac{1}{82}$ of an inch; greatest breadth $\frac{1}{330}$.

Tab. XXVII. fig. 1. *a.* front view; *b.* side view.

2. *C. acerosum* (Schrank); frond linear-lanceolate, gradually tapering; ends conical; fillets distinct; vesicles in a single series; empty frond colourless; striæ none or indistinct.

β. Frond more elongated; striæ more distinct.

Vibrio acerosus, Schrank, *Faun. Boica,* 3. 2. p. 47 (1803).
Closterium acerosum, Ehrenberg, *Abhandl. der Berl. Akademie* (1831); *Infus.* p. 92. t. 6. f. 1. Meneghini, *Synop. Desmid. in Linnæa* 1840, p. 233. Kützing, *Phycologia Germanica,* p. 131. Jenner, *Fl. of Tunbridge Wells,* p. 196. Hassall, *British Freshwater Algæ,* p. 374.

Carnarvon; Dolgelley; and Penzance, *J. R.* Several stations in Sussex, and near Reigate, Surrey, *Mr. Jenner.* Cheshire and Westmoreland, *Mr. Sidebotham.* Manchester, *Mr. Williamson.* Aberdeenshire and Banffshire, *Mr. P. Grant.* Bristol, *Mr. Thwaites.*

β. Brackish water at Shirehampton, near Bristol, *Mr. Thwaites.* Germany, *Ehrenberg, Kützing.* Falaise, *Brébisson.* Mexico, *Bailey.*

Frond bright green, slender, six to twelve times longer than broad, linear-lanceolate; ends conical; lower margin nearly straight, except at the extremities, the conical form of which causes it to incline upwards; the upper margin slightly convex. Fillets three or more; vesicles arranged in a single longitudinal series.

The empty frond is colourless, and has a transverse line at the suture. This species is generally described as destitute of striæ; but Mr. Jenner and myself have frequently detected them, although seldom distinctly. The striæ are more evident in some specimens of β.

Sporangia have been gathered by Mr. Jenner in Sussex; they are orbicular, and placed between the deciduous fronds.

Closterium acerosum differs from *C. Lunula* and *C. lanceolatum* in its more slender frond. Some states of it approach nearer to *C. turgidum*; it is usually however more slender, and its empty frond also is colourless, and its striæ, whenever they occur, are very indistinct.

Length of frond from $\frac{1}{170}$ to $\frac{1}{58}$ of an inch; greatest breadth from $\frac{1}{1103}$ to $\frac{1}{582}$; diameter of sporangium $\frac{1}{408}$.

Length of variety β. from $\frac{1}{48}$ to $\frac{1}{32}$; greatest breadth $\frac{1}{510}$.

Tab. XXVII. fig. 2. *a.* front view of frond with endochrome; *e, f, g.* small states; *b, d.* empty fronds; *h, i.* conjugating fronds; *k, l.* perfect sporangia; *m.* germinated capsule?; *c.* frond of β.

3. *C. lanceolatum* (Kütz.); frond semilanceolate, gradually tapering; ends subacute; fillets several; vesicles in a single series; empty frond colourless.

Cymbella Hopkirkii, Moore, *in Harvey's Manual of British Algæ,* p. 215 (scarcely *Conferva ovalis,* Hopkirk.) (1841).

Closterium tenue, Bailey, *American Journal of Science and Arts*, v. 41. p. 303. t. 1. f. 37 (1841)?
Closterium lanceolatum, Kützing, *Phycologia Germanica*, p. 130 (1845), *in lit. cum icone*. Brébisson, *in lit.*

About Lisburn; and near Belfast, *Mr. D. Moore.* Galway, *Mr. M'Calla.* Ambleside, *Mr. Sidebotham.* Westerham, Kent, *Mr. Jenner.*

West Point, New York, *Bailey.* Germany, *Kützing.* Falaise, *Brébisson.*

Frond semilanceolate, six to ten times longer than broad; stouter than that of *Closterium acerosum*; lower margin straight, but inclined upwards at each end owing to the tapering of the extremities; upper margin convex; fillets three or more; vesicles in a single longitudinal series. Empty frond usually colourless and destitute of striæ.

Closterium lanceolatum is larger than *C. acerosum*; instead of being nearly alike, its margins, as stated above, differ considerably from each other, and its extremities are more tapering. It is distinguished from *C. Lunula* by having its vesicles in a single series, and its upper margin much less convex. Its ends are neither curved upwards, nor so obtuse as those of *C. turgidum*.

Length of frond $\frac{1}{64}$ of an inch; greatest breadth $\frac{1}{453}$.

Tab. XXVIII. fig. 1. *a.* frond with endochrome; *b.* empty frond.

4. **C. turgidum** (Ehr.); lower margin of frond slightly concave, inclined upwards at the rounded ends; upper margin with a depression near each extremity; empty frond coloured; striæ numerous, fine but distinct.

Closterium turgidum, Ehrenberg, *Infus.* p. 95. t. 6. f. 7 (1838). Meneghini, *Synopsis Desmid. in Linnæa* 1840, p. 234. Jenner, *Fl. of Tunbridge Wells*, p. 196. Kützing, *Phycologia Germanica*, p. 131. Hassall, *British Freshwater Algæ*, p. 371.

Carnarvon; Dolgelley; Penzance, *J. R.* Sussex; Kent; Surrey; Hampshire, *Mr. Jenner.* Aberdeenshire, *Dr. Dickie* and *Mr. P. Grant.* Cheshire and Westmoreland, *Mr. Sidebotham.*

Germany, *Ehrenberg.* France, *Brébisson.*

Frond stout, visible to the naked eye, green, six to ten times longer than broad, semilanceolate; extremities curved upwards, ends broadly rounded. The lower margin is somewhat concave, but always sloped upwards at each end, where the upper margin, which is convex, has a slight depression in consequence of the inclination of the apex upwards. The vesicles form a single longitudinal series; fillets three or more.

The empty frond is reddish or straw-colour, and somewhat opake; the suture at the middle is distinct, and so are the longitudinal striæ, which are numerous, close and fine.

The curved and rounded ends are characteristic of this species. *Closterium decussatum*, Kütz., differs in its more attenuated extremities.

Length of frond $\frac{1}{39}$ of an inch; greatest breadth $\frac{1}{370}$; distance between the striæ $\frac{1}{27200}$.

Tab. XXVII. fig. 3. *a.* frond with endochrome; *b.* empty fronds.

†† *Frond smooth, crescent-shaped, rapidly tapering from the middle.*

5. *C. Ehrenbergii* (Menegh.); frond smooth, crescent-shaped, when empty colourless; lower margin inflated at the middle; ends rounded; vesicles numerous, scattered.

Closterium Lunula, Ehr. *Infus.* t. 5. f. xv. 2 (1838). Hassall, *Brit. Freshwater Algæ*, t. 84. f. 4?
Closterium Ehrenbergii, Meneghini, *Synop. Desmid. in Linnæa* 1840, p. 232. Hassall, *Brit. Algæ*, p. 369 (excl. synonyms), t. 87. f. 1.

Common, often in streams. Henfield, Westham, Tunbridge Wells, &c., Sussex; Chertsey and Reigate, Surrey, *Mr. Jenner.* Penzance and Dolgelley, *J. R.* Bristol, *Mr. Thwaites.* Ambleside, Westmoreland, *Mr. Sidebotham.* Aberdeenshire and Banffshire, *Mr. P. Grant.*

Germany, *Ehrenberg.* Falaise, *Brébisson.*

Frond bright green, stout, distinctly visible to the naked eye, lunately curved, five or six times longer than broad, extremities tapering, ends rounded, lower margin very concave with an inflation at the middle; fillets several, three of them more evident; vesicles numerous, small and scattered. Empty frond colourless, without a suture at the middle, collapsing when dried.

Closterium Ehrenbergii agrees with *C. Lunula* and *C. moniliferum* in size, colour and texture: it is more curved than the former, the lower margin being concave, not straight, and its centre is protuberant. From the latter it differs in its vesicles, which are smaller, more numerous and scattered.

Length of frond $\frac{1}{68}$ of an inch; greatest breadth $\frac{1}{400}$.

Tab. XXVIII. fig. 2. frond with endochrome.

6. *C. moniliferum* (Bory); frond smooth, crescent-shaped, when empty colourless; lower margin inflated at the middle; ends rounded; vesicles in a single row.

Lunulina monilifera, Bory, *Encycl. Méthod. Hist. N. des Zooph.* 1824, t. 3. f. 22, 25 and 27.
Closterium Lunula, var., Ehr. *Abh. der Berl. Ak.* 1830, p. 62.
Closterium acerosum, var., Ehr. *Abh. der Berl. Ak.* 1831, p. 68.
Closterium Lunula, Kützing, *Synop. Desmid. in Linnæa* 1833, f. 80. Brébisson, *Alg. Fal.* p. 58. t. 8.
Closterium moniliferum, Ehr. *Infus.* p. 90. t. 5. f. 16 (1838). Meneghini, *Synop. Desmid. in Linnæa* 1840, p. 232. Bailey, *Amer. Bacil. in Amer. Journ. of Science and Arts*, v. 41. p. 302. t. 1. f. 31. Jenner, *Fl. of Tunbridge Wells*, p. 196. Kützing, *Phy. Germ.* p. 130. Hassall, *Brit. Alg.* p. 370.

Common. Penzance and Dolgelley, *J. R.* Framfield and Hastings, Sussex; and Reigate, Surrey, *Mr. Jenner.* Kerry, *Mr. Andrews.* Cheshunt, *Mr. Hassall.* Near Bristol, *Mr. Thwaites.* Yate, near Bristol, *Mr. Broome.* Aberdeenshire, *Dr. Dickie* and *Mr. P. Grant.* Banffshire, *Mr. P. Grant.* Cheshire and Westmoreland, *Mr. Sidebotham.* Manchester, *Mr. Williamson.*

Germany, *Ehrenberg, Kützing, Corda,* &c. Falaise, *Brébisson.* Common in New York and New England, *Bailey.*

Frond bright green, stout, distinctly visible to the naked eye, lunately curved, five or six times longer than broad, extremities tapering, ends rounded; lower margin very concave, inflated at the middle; fillets several, one to three more distinct than the rest; vesicles few, large, and disposed in a single longitudinal series. The empty frond, colourless and without a suture at the middle, collapsing when dried.

Closterium moniliferum is smaller than *C. Lunula* and more curved, and it may always be distinguished by the inflation at the middle of its lower margin; its vesicles form a single series, a character which separates it also from *C. Ehrenbergii*.

Length of frond from $\frac{1}{75}$ to $\frac{1}{60}$ of an inch; greatest breadth from $\frac{1}{510}$ to $\frac{1}{466}$.

Tab. XXVIII. fig. 3. *a.* frond with endochrome; *b.* empty frond.

7. *C. Jenneri* ——; frond crescent-shaped, generally slightly constricted at the suture, when empty colourless, rapidly tapering; ends very obtuse; vesicles in a single series.

Closterium moniliferum, Ehr. *Infus.* t. 5. f. 16. n. 6, 7 (1838).

Sussex; Surrey; and Hampshire, *Mr. Jenner*. Penzance, *J. R.*

Frond smaller than that of *Closterium moniliferum*, much curved; extremities tapering; ends rounded. There is generally a slight constriction at the suture. Fillets two or three. The empty frond colourless and without markings.

Closterium Jenneri is more curved than *C. moniliferum*, and there is no inflation of its lower margin. It is more curved than *C. Leibleinii* and *C. Dianæ*, and its ends are far more obtuse than theirs.

Length between ends of frond $\frac{1}{281}$ of an inch; breadth at suture $\frac{1}{1730}$.

Tab. XXVIII. fig. 6. *a, b.* fronds with endochrome; *c.* empty frond.

8. *C. Leibleinii* (Kütz.); frond smooth, crescent-shaped; extremities much attenuated and subacute at the apex; lower margin slightly inflated at the middle; vesicles in a single row.

β. More slender, the central inflation less evident; empty frond of a deeper colour and its central suture distinct.

Closterium Lunula, Leibl. *Fl.* 1827, p. 259. according to *Kütz.*
Closterium Leibleinii, Kützing, *Synop. Diatom. in Linnæa* 1833, p. 596; Phycol. *Germ.* p. 130. Brébisson, *Alg. Fal.* p. 58. t. 8. Meneghini, *Synop. Desmid. in Linnæa* 1840, p. 232.

Cheshire, *Mr. Sidebotham*. Albourn, Framfield, and Hastings, Sussex; and Reigate, Surrey, *Mr. Jenner*. Penzance and Dolgelley, *J. R.* Rochdale, *Mr. Coates*. Aberdeen, *Mr. P. Grant*. Shirehampton near Bristol, *Mr. Thwaites*.

Germany, *Kützing*. Falaise, *Brébisson*.

Frond small, crescent-shaped, stout, about four to eight times as long as broad; extremities tapering to a subacute apex. At the centre of the lower margin is a slight protuberance, which however is often nearly obsolete. Fillets frequently obscure; vesicles few, large, and disposed in a single series. The empty frond has generally a pale straw-colour tinge, and a transverse suture is usually visible at the centre.

Closterium Leibleinii much resembles a young specimen of *C. moniliferum*; but it is smaller and more curved, the ends are more acute, and the empty frond is usually coloured.

The variety β. is nearly intermediate between this and the following species.

I have gathered the sporangia at Penzance and Dolgelley; Mr. Jenner has found them in Sussex, and Mr. Thwaites near Bristol. They are orbicular, and but loosely connected with the empty segments of the fronds.

Length of frond from $\frac{1}{291}$ to $\frac{1}{165}$ of an inch; greatest breadth from $\frac{1}{1032}$ to $\frac{1}{582}$; length of sporangium from $\frac{1}{1360}$ to $\frac{1}{942}$; breadth from $\frac{1}{2040}$ to $\frac{1}{1275}$.

Tab. XXVIII. fig. 4. *a, c, f, i, k.* different states of frond with endochrome; *b, d, e.* empty fronds; *g, h, l.* sporangia.

9. *C. Dianæ* (Ehr.); fronds smooth, slender-crescent-shaped; extremities tapering; apex subacute; lower margin not inflated; vesicles in a single series.

Closterium ruficeps, Ehr. *Abh. der Berl. Ak.* 1831, p. 67.
Closterium Dianæ, Ehr. *Infus.* p. 92. t. 5. f. 17 (1838). Kützing, *Phy. Germ.* p. 130. Hassall, *Brit. Freshwater Alg.* p. 371.

Sussex, *Mr. Jenner.* Dolgelley, *J. R.* Aberdeenshire, *Dr. Dickie* and *Mr. P. Grant.* Cheshire and Ambleside, *Mr. Sidebotham.*

Germany, *Ehrenberg.* Falaise, *Brébisson.*

Frond slender, six to eight times longer than broad, curved; extremities tapering to a subacute point. The lower margin is concave and not inflated at the centre. Fillets obscure; vesicles disposed in a single longitudinal series.

The empty frond has a pale straw-colour tint, and a transverse suture at the centre, where, as in *C. Leibleinii*, the segments easily separate.

This plant differs from *C. Leibleinii* in being longer, more slender and less curved, and in having no projection at the centre of the lower margin; yet I have doubts whether they are really distinct, as I have seen some specimens apparently intermediate.

Length between ends of frond $\frac{1}{143}$ of an inch; greatest breadth $\frac{1}{1275}$.

Tab. XXVIII. fig. 5. *a, b.* fronds with endochrome; *c.* empty frond.

††† *Frond nearly straight, scarcely attenuated; ends truncate; longitudinal striæ none or indistinct.*

10. *C. didymotocum* (Corda); frond nearly straight, broadly linear; extremities slightly tapering; ends truncate, reddish; fillets obscure.

α. Empty frond divided into four portions by three transverse lines or sutures.

Closterium didymotocum, Corda, *Almanach de Carlsbad*, 1835, p. 125. t. 5. f. 64, 65. Brébisson, *in lit. cum icone et specimine*.
Closterium subrectum, Brébisson, *Alg. Fal.* p. 59. t. 8 (1835). Kützing, *Phycologia Germanica*, p. 131.

β. *Baillyanum* (Bréb.); frond smaller with a suture only at the middle.

Closterium Baillyanum, Brébisson, *in lit. cum icone et specimine* (1845).

α. Rare. Dolgelley, *J. R.*

β. Common. Dolgelley and Penzance, *J. R.* Sussex; Surrey; and Hampshire, *Mr. Jenner.* Aberdeenshire, *Mr. P. Grant.* Ambleside, *Mr. Sidebotham.*

Germany, *Corda, Kützing.* α. et β. Falaise, *Brébisson.*

Frond six to ten times longer than broad, stout, nearly straight, equal except at the extremities, which taper more or less; ends truncate, fillets obscure; vesicles in a single longitudinal series; terminal globules and moving granules distinct.

Empty frond reddish, especially at the ends, where the colour is conspicuous even before the endochrome has collapsed. Faint striæ may generally be detected in old specimens.

The variety β. is by M. de Brébisson considered a distinct species. The only difference I can discover is in the number of sutures, and I doubt whether any dependence can be placed on this character. *Closterium Baillyanum* is also usually smaller than α, but not invariably, and before the escape of the endochrome I am unable to distinguish them even as varieties.

Closterium didymotocum may be known from all the preceding species by its straight frond combined with truncate ends. It differs from *C. striolatum*, *C. intermedium*, and *C. angustatum*, in the absence or indistinctness of its striæ and the upward inclination of the lower margin at the extremities.

Length of frond in β. $\frac{1}{65}$ of an inch; breadth $\frac{1}{813}$.

Tab. XXVIII. fig. 7. *a.* frond of α. with endochrome; *b.* empty frond; *c.* frond of β. with endochrome; *d.* empty frond.

†††† *Empty frond distinctly striated, mostly coloured.*

11. *C. attenuatum* (Ehr.); frond curved, attenuated, suddenly contracted at the end into a conical point; empty frond reddish, faintly striated.

Closterium attenuatum, Ehrenberg, *Infusor.* p. 94. t. 6. f. 4 (1838). Kützing, *Phycologia Germanica*, p. 131. Brébisson, *in lit. cum icone.*

Chiltington Common near Pulborough; Midhurst; Ashdown Forest; near Tunbridge Wells, &c., Sussex; Reigate, Surrey, *Mr. Jenner.* Dolgelley and Penzance, *J. R.* Ambleside, *Mr. Sidebotham.*

Berlin, *Ehrenberg.* Falaise, *Brébisson.*

Frond rather larger than that of *Closterium striolatum*, eight to twelve times longer than broad, curved, gradually attenuated, and at the end suddenly contracted into an obtuse, conical or subcylindrical point. Fillets obscure; vesicles in a single longitudinal series. Empty frond reddish, especially at the ends, transverse suture distinct. Longitudinal striæ numerous, close, generally faint, but sometimes very distinct.

Closterium attenuatum might be supposed an incomplete state of another species, especially when its ends are dissimilar; but the sudden contraction of the extremities is a sufficient distinction.

Length of frond $\frac{1}{57}$ of an inch; greatest breadth $\frac{1}{669}$; distance between the striæ $\frac{1}{26315}$.

Tab. XXIX. fig. 5. *a.* frond with endochrome; *b.* empty frond.

12. **C. costatum** (Corda); frond stout, semilunate or crescent-shaped, tapering from the middle; ends obtuse; striæ few and conspicuous; suture solitary.

Closterium costatum, Corda, *Almanach de Carlsbad*, 1835, p. 124. t. 5. f. 61 to 63; *Observations microscopiques sur les Animalcules des Eaux et des Thermes de Carlsbad*, p. 34. Brébisson, *in lit. cum icone.*
Closterium doliolatum, Brébisson, Meneghini, *Synopsis Desmid. in Linnæa* 1840, p. 237.
Closterium dilatatum, Kützing, *Phycologia Germanica*, p. 132 (1845); *in lit. cum icone.*

Dolgelley; and Penzance, *J. R.* Sussex; Surrey; and Hampshire, *Mr. Jenner.* Aberdeenshire, *Dr. Dickie* and *Mr. P. Grant.* Ambleside, *Mr. Sidebotham.* Near Bristol, *Mr. Broome.*

Carlsbad; and Prague, *Corda.* Falaise, *Brébisson.*

Frond stout, five or six times longer than broad, crescent-shaped or nearly semilunate, rapidly attenuated from the centre; end obtuse or somewhat truncate. Fillets obscure; vesicles in a single series.

The empty frond is reddish, with a single transverse suture generally of two or more lines. The longitudinal striæ, which are few compared to the breadth of the frond, are so very distinct that they may be counted without difficulty.

I have gathered a single sporangium at Dolgelley; it was orbicular, and resembled that of *Closterium striolatum.*

This species agrees with *Closterium angustatum* in its distinct, subdistant striæ; but their forms are widely different. It is stouter in proportion to its length than either *C. turgidum* or *C. striolatum*, and its striæ are much fewer than theirs.

Length of frond $\frac{1}{75}$ of an inch; greatest breadth $\frac{1}{384}$; distance of striæ apart $\frac{1}{5319}$.

Tab. XXIX. fig. 1. *a.* frond with endochrome; *b.* empty frond.

13. **C. striolatum** (Ehr.); frond closely but distinctly striated, semilunate or crescent-shaped, tapering from the middle; sutures generally three, never more.

Closterium striolatum, Ehrenberg, *Abh. der Berlin. Ak.* 1833, p. 68; *Infusor.*

p. 95. t. 6. f. 12. Meneghini, *Synopsis Desmid. in Linnæa* 1840, p. 234. Bailey, *Amer. Journ. of Science and Arts*, v. 41. p. 303. t. 1. f. 35. Kützing, *Phycologia Germanica*, p. 131. Jenner, *Fl. of Tunbridge Wells*, p. 196. Hassall, *Brit. Freshwater Algæ*, p. 373.

Common. Carnarvon; Dolgelley; Tal Sarn near Lampeter; near Carmarthen; and Penzance, *J. R.* Sussex; Surrey; Kent; and Hampshire, *Mr. Jenner.* Kerry, *Mr. Andrews.* Galway, *Mr. M'Calla.* Hertfordshire, *Mr. Hassall.* Aberdeenshire, *Dr. Dickie* and *Mr. P. Grant.* Cheshire and Westmoreland, *Mr. Sidebotham.* Manchester, *Mr. Gray* and *Mr. Williamson.*

Germany, *Ehrenberg.* New York; and New England, *Bailey.* Falaise, *Brébisson.*

Frond six to twelve times longer than broad, very variable in both length and breadth, curved, tapering from the middle; ends very obtuse; upper margin convex; lower concave or straight, but never inclined upwards at the ends. Vesicles in a single series; fillets often obscure. This species forms yellowish-brown masses in the water.

The empty frond is reddish and darkest at the ends. Striæ numerous, crowded, and easily detected. Sutures usually three, but sometimes only two: specimens with a single suture are very rare.

The sporangia, which are not uncommonly met with, are orbicular and placed between the deciduous fronds.

Closterium striolatum is shorter than *C. turgidum*, and the ends do not turn up; its striæ are much more numerous and close than those of *C. costatum*; and both *C. turgidum* and *C. costatum* have constantly but a single suture. It is more closely striated than either *C. intermedium* or *C. angustatum*, and stouter also in proportion to its length. With this species *C. didymotocum* in some respects agrees as to its size and form, but it is straighter and less tapering, and its ends are more truncate, and the striæ, if present, are detected with the greatest difficulty.

Length of frond from $\frac{1}{80}$ to $\frac{1}{68}$ of an inch; greatest breadth from $\frac{1}{625}$ to $\frac{1}{535}$; distance between the striæ $\frac{1}{10,000}$.

Tab. XXIX. fig. 2. *a, b, c.* fronds with endochrome; *d, e, f.* empty fronds; *g, h.* sporangia.

14. *C. intermedium* ——; frond slender, slightly curved, tapering; striæ distinct, not crowded; sutures usually more than three.

Dolgelley, *J. R.*

Frond slender, nearly straight, many times longer than broad, very gradually tapering, ends truncate. Fillets obscure; vesicles in a single series. This species forms in the water large masses of a yellow-brown colour.

The empty frond is straw-colour; its striæ are distinct and can be counted without difficulty. Sutures, though occasionally but three, are usually from four to seven in number. The central suture often forms a pellucid line resembling a dissepiment.

In its form and in the number of its striæ *Closterium intermedium* is intermediate between *C. striolatum* and *C. angustatum*. According to M. de Brébisson, *C. striolatum* has about forty striæ, *C. intermedium* twenty, and *C. an-*

gustatum ten or twelve. The most remarkable feature in this plant is the number of its sutures exceeding that of any other species; it is more slender and tapers more gradually than *C. striolatum*. Its more numerous striæ and tapering form distinguish it from *C. angustatum*.

Length of frond from $\frac{1}{77}$ to $\frac{1}{54}$ of an inch; greatest breadth $\frac{1}{1073}$; distance between the striæ $\frac{1}{14571}$.

Tab. XXIX. fig. 3. *a.* frond with endochrome; *b, c.* empty fronds.

15. *C. angustatum* (Kütz.); frond sublinear, curved, scarcely attenuated; ends truncate; striæ few, very distinct and prominent; sutures usually three.

Closterium angustatum, Kützing, *Phycologia Germanica*, p. 132 (1845); *in lit. cum icone.*
Closterium sulcatum, Brébisson, *in lit. cum icone* (1845).

Dolgelley, *J. R.* Waterdown Forest, &c., Sussex; Reigate, Surrey, *Mr. Jenner.* Aberdeenshire, *Mr. P. Grant.* Ambleside, *Mr. Sidebotham.*

Germany, *Kützing.* Falaise, *Brébisson.*

Frond curved, eight or more times longer than broad, nearly equal in breadth except towards the extremities which are somewhat attenuated, their ends however being truncate. The vesicles are arranged in a single longitudinal series; fillets obscure.

The empty frond is of a pale reddish colour and darkest at the ends. Transverse sutures usually three, the central one of two or more lines. The longitudinal striæ, of which seldom more than three or four can be seen at one view, are few in number, very distinct and prominent, and not unfrequently somewhat spiral. The frond, I believe, has several angles or ridges, which under the glass have the effect of striæ.

Its linear form and scarcely attenuated extremities distinguish this species from *Closterium intermedium*. The striæ also are fewer in number, less crowded, and more distinct. *C. angustatum* differs from *C. juncidum* by its stouter form and more distinct striæ; its upper margin also is convex, and not straight as in that species.

Length of frond $\frac{1}{60}$ of an inch; greatest breadth $\frac{1}{1142}$; distance between the striæ $\frac{1}{5000}$.

Tab. XXIX. fig. 4. *a.* frond with endochrome; *b, c.* empty fronds.

16. *C. juncidum* ———; frond elongated, very slender, linear, straight except at the extremities, which are slightly attenuated and curved downwards.

β. Frond stouter and less elongated.

α. Dolgelley and Penzance, *J. R.* Chiltington Common near Pulborough, Sussex, *Mr. Jenner.* Aberdeenshire, *Dr. Dickie* and *Mr. P. Grant.* Ambleside, Westmoreland; and Congleton, Cheshire, *Mr. Sidebotham.*

β. Dolgelley, *J. R.* Midhurst, Sussex, *Mr. Jenner.*
Falaise, *Brébisson.*

Frond many times longer than broad, very slender, linear, straight except at the extremities, which are slightly attenuated and curved downwards; ends obtuse. Fillets and vesicles none or obscure.

In α. the empty frond is nearly colourless, and the longitudinal striæ are faint. In β. the colour is deeper and the striæ more conspicuous. Transverse sutures usually three.

I have gathered sporangia at Dolgelley and Penzance; they are orbicular and placed between the deciduous fronds.

The slender linear frond distinguishes this species.

Length of frond from $\frac{1}{111}$ to $\frac{1}{69}$ of an inch; breadth $\frac{1}{5000}$; diameter of sporangium $\frac{1}{672}$; distance between the striæ $\frac{1}{40000}$.

Length of β. $\frac{1}{144}$; breadth $\frac{1}{2040}$; length of sporangium $\frac{1}{625}$; breadth $\frac{1}{688}$; breadth between the striæ $\frac{1}{27777}$.

Tab. XXIX. fig. 6. *a.* frond of α. with endochrome; *b.* empty frond; *c, d.* sporangia. Fig. 7. *a.* frond of β. with endochrome; *b.* empty frond; *c, d.* sporangia.

** *Frond striated, much elongated, gradually tapering, scarcely rostrate; sporangium bilobed, situated between the conjugated fronds.*

17. *C. lineatum* (Ehr.); frond striated, slender, long, curved, gradually tapering into the conico-rostrate extremities; lower margin slightly inflated at its centre.

β. Longitudinal striæ spiral.

Closterium lineatum, Ehrenberg, *Abhandl. der Berlin. Akademie* 1833, p. 238;
    Infusor. p. 95. t. 6. f. 8. Meneghini, *Synop. Desmid. in Linnæa* 1840, p. 234. Kützing, *Phycologia Germanica*, p. 131. Jenner, *Fl. of Tunbridge Wells*, p. 196. Hassall, *British Freshwater Algæ*, p. 372.
Closterium elongatum, Brébisson, Meneghini, *Synopsis Desmid.* p. 234 (1840).

Carnarvon; Dolgelley; and Penzance, *J. R.* Ashdown Forest, and Waterdown Forest, Sussex, *Mr. Jenner.* Aberdeenshire; and Banffshire, *Mr. P. Grant.* Manchester, *Mr. Sidebotham.*

Germany, *Ehrenberg, Kützing.* Falaise, *Brébisson.* Mexico, *Bailey.*

Frond large, twelve or more times longer than broad, slightly curved, lanceolate, gradually tapering into the slender extremities, which are curved downwards and obtuse at the end. The upper margin is slightly but uninterruptedly convex, so as to form an arc; the lower margin is concave, owing to the curved extremities, but it is slightly protuberant at the centre. Endochrome reaching nearly to the ends; fillets three or more, frequently obscure; vesicles in a single longitudinal series.

Empty frond of a pale straw-colour, with numerous but distinct longitudinal striæ, and one or more transverse lines at the centre. Specimens occasionally are met with having the striæ spirally arranged. This state, which is very beautiful, I can only regard as an accidental variety, as I have observed such a frond conjugated with one of the common kind.

Closterium lineatum connects the rostrate with the other striated species, as in general aspect it agrees with some of the latter, and in its inflated centre and tapering extremities with the former. It is liable to be mistaken for *C. Ralfsii*, but it is longer, more slender, less inflated at the centre, and more gradually attenuated; the colour too of its empty frond is paler, and its striæ are not quite so crowded.

I have frequently gathered the sporangium near Dolgelley, and Mr. Jenner has gathered it in Sussex. It is very remarkable, being the only known example of a geminate or bilobed sporangium. The fronds approach and couple in the usual manner; but instead of the contents of both fronds uniting in the ordinary manner into a single body, a bilobed body is produced not unlike a species of *Cosmarium* in form. A distinct line is perceptible between the lobes or portions, and each resembles a globular or oval body flattened at the junction, and is formed by the contents of the adjacent segments. The sporangium is more closely connected with the conjugated fronds than is the case in species belonging to the first section, but the fronds are not permanently united to it as in *Closterium rostratum*. Although I have here called it bilobed, I regard the sporangium as binate, rather than bilobed, because it readily separates at the junction, each portion retaining the segments of the frond belonging to it.

Length of frond $\frac{1}{48}$ of an inch; greatest breadth $\frac{1}{909}$; length of the bilobed sporangium $\frac{1}{173}$; greatest breadth $\frac{1}{370}$; distance between the striæ $\frac{1}{26315}$.

Tab. XXX. fig. 1. *a.* frond with endochrome; *b.* empty frond; *c.* sporangium.

*** *Frond either rostrate or minute, colourless and acicular; sporangium cruciform.* (Stauroceras, *Kütz.*)

† *Frond striated, tapering at each end into a distinct beak.*

18. *C. Ralfsii* (Bréb.); frond stout, striated, curved, rapidly attenuated into linear beaks which are shorter than the ventricose body.

Closterium rostratum, Ralfs, *in Jenner, Fl. of Tunbridge Wells*, p. 196 (1845) (not of Ehrenberg according to Brébisson).
Closterium Ralfsii, Brébisson, *in lit.* (1845).
Stauroceras Ralfsii, Brébisson, *in lit.* (1846).

Carnarvon and Dolgelley, *J. R.* Midhurst; near Tunbridge Wells, &c., *Mr. Jenner.* Aberdeenshire, *Dr. Dickie.* Near Manchester; and Ambleside, *Mr. Sidebotham.*

Falaise, *Brébisson.*

Frond yellowish-brown, stout, six to eight times longer than broad; the upper margin convex, the lower concave, but ventricose at the centre; extremities tapering into a narrow linear beak, which is curved downwards, shorter than the body, and obtuse at the apex. Vesicles disposed rather irregularly in a single longitudinal series. Fillets generally obscure.

The empty frond, which is firm, is reddish, especially at the ends. Striæ numerous, close, and distinct; transverse suture solitary.

This and the two following plants differ considerably from the other striated species (*Closterium lineatum* in some respects excepted). The body of the frond is somewhat lanceolate, being equally convex on both margins, and tapering at each end into a linear or setaceous beak, which is curved downward and thus gives the frond a curved appearance. The green endochrome is nearly confined to the inflated body.

Closterium Ralfsii may be known from *C. rostratum* and *C. setaceum* by its larger size, the deeper colour of its empty frond, the shortness of its beaks, its firmer texture, and by the upper margin, which, inclusive of the beaks, forms an uninterrupted convexity or arc.

Length of frond $\frac{1}{79}$ of an inch; greatest breadth $\frac{1}{526}$; distance between the striæ $\frac{1}{17241}$.

Tab. XXX. fig. 2. *a.* frond with endochrome; *b.* empty frond.

19. *C. rostratum* (Ehr.); frond striated, tapering at each end into a setaceous curved beak which is about equal in length to the ventricose body; sporangium cruciform.

Closterium rostratum, Ehrenberg, *Abhandl. der Akademie d. Wissensch. zu Berlin* 1831, p. 67; 1833, p. 240; *Infus.* p. 97. t. 6. f. 10. Meneghini, *Synop. Desmid. in Linnæa* 1840, p. 235.
Closterium Acus, Nitzsch, Kützing, *Synopsis Diatomearum in Linnæa* 1833, p. 595. f. 81; *Alg. aq. dulc.* No. 80. Jenner, *Fl. of Tunbridge Wells*, p. 196.
Closterium caudatum, Corda, *Almanach de Carlsbad* 1835, p. 125. t. 5. f. 66.
Stauroceras Acus, Kützing, *Phycologia Germanica*, p. 133 (1845). Brébisson, *in lit. cum specimine.*

Dolgelley and Penzance, *J. R.* Sussex; Kent, *Mr. Jenner.* Durdham Down near Bristol, *Mr. Thwaites.* Near Congleton, Cheshire; Ambleside, Westmoreland; and near Manchester, *Mr. Sidebotham.* Aberdeenshire, *Mr. P. Grant.*

Germany, *Ehrenberg, Kützing, Corda.* Falaise, *Brébisson.*

Frond lanceolate, tapering into setaceous beaks, which are curved downwards at the extremity, obtuse at the apex, and nearly equal in length to the inflated body.

Endochrome green, confined to the inflated portion; fillets obscure; vesicles arranged in a single series; the moving granules are situated at the end of the endochrome, and apparently are not contained within a globule, as their motion has a wider range than in many other species.

The empty frond is colourless or tinged straw-colour, and the striæ are numerous and close.

Mr. Jenner and myself have gathered conjugated specimens of *Closterium rostratum* in greater abundance than those of any other species. The cell formed for the reception of the sporangium is intimately connected with the fronds, of which indeed it seems to form a part. When viewed in front it may be described as a four-sided figure, whose angles are cut off at their union with the segments. All its sides are concave, but those between the segments of the same frond are shorter than the others and have deeper and more angular concavities, so that the sporangium has the same figure as the cell.

Length of frond $\frac{1}{69}$ of an inch; greatest breadth $\frac{1}{1068}$; length of sporangium $\frac{1}{114}$; breadth $\frac{1}{346}$; distance between the striæ $\frac{1}{25000}$.

Tab. XXX. fig. 3. *a.* frond with endochrome; *b, d.* empty fronds; *c.* conjugated fronds; *e.* sporangium.

20. *C. setaceum* (Ehr.); frond very slender, finely striated, narrow-lanceolate, tapering at each extremity into a very long setaceous beak, which alone is curved; vesicles none or obscure.

Closterium setaceum, Ehrenberg, *Abhandl. der Berlin. Ak.* 1833, p. 239; *Infus.* p. 97. t. 6. f. 11. Meneghini, *Synop. Desmid.* p. 235. Jenner, *Fl. of Tunbridge Wells*, p. 196. Hassall, *Brit. Freshwater Algæ*, p. 373.
Closterium rostratum, Bailey, *Amer. Bacillaria in Amer. Journ. of Science and Arts*, v. 41. p. 303. t. 1. f. 36 (1841).
Stauroceras subulatum, Kützing, *Phycologia Germanica*, p. 133 (1845).
Stauroceras setaceum, Brébisson, *in lit.* (1846).

Dolgelley, *J. R.* Waterdown Forest; and near Cross-in-hand, Sussex, *Mr. Jenner.* Near Bristol, *Mr. Broome.* Near Aberdeen, *Mr. P. Grant.*

Germany, *Ehrenberg.* Staten Island, New York, *Bailey.* Falaise, *Brébisson.*

Frond minute, many times longer than broad, very slender; the body narrow-lanceolate, straight, attenuated at each end into a setaceous beak or awn, which is longer than the inflated portion, curved downwards at the extremity and blunt at the apex. The endochrome is pale, and does not extend beyond the inflated part. Vesicles none or indistinct; fillets none; moving granules not contained within a globule.

The empty frond is colourless, and exhibits close and faint striæ.

Conjugated specimens are not uncommon. The process differs but little from that described under the preceding species. The sporangium in this species also appears like a continuation of the segments, and is quadrate or cruciform.

Closterium setaceum may be known from *C. rostratum* and *C. gracile* by its slender beaks being longer than the body of the frond.

Length of frond $\frac{1}{116}$ of an inch; greatest breadth $\frac{1}{2381}$; distance between the striæ $\frac{1}{17857}$.

Tab. XXX. fig. 4. *a.* frond with endochrome; *b.* empty frond; *c.* sporangium.

†† *Frond minute, tapering, not rostrate; empty frond colourless and without markings.*

21. *C. Cornu* (Ehr.); frond smooth, minute, curved, very slender; extremities slightly attenuated; ends obtuse; vesicles none or indistinct.

β. Frond more turgid.

Closterium Cornu, Ehrenberg, *Abhandl. der Berlin. Ak.* p. 62 (1830); *Infusor.* p. 94. t. 6. f. 5. Meneghini, *Synop. Desmid. in Linnæa* 1840, p. 233. Jenner, *Fl. of Tunbridge Wells*, p. 196. Kützing, *Phycologia Germanica*, p. 131. Hassall, *Brit. Freshwater Algæ*, p. 372.
Closterium tenue, Kützing, *Synopsis Diatom. in Linnæa* 1833, p. 595. f. 78; *Phycologia Germanica*, p. 130.

Henfield, Piltdown Common, Battle, and Rackham Bogs, Sussex, *Mr. Jenner.* Dolgelley, *J. R.* Ambleside, *Mr. Sidebotham.*

Germany, *Ehrenberg, Kützing.* Falaise, *Brébisson.*

Frond very minute, five to eight times longer than broad, slender, slightly curved; extremities attenuated; ends blunt, but frequently one more so than the other. The figure of the frond is variable; the upper margin is more or less convex, the lower one concave or straight; occasionally the extremities curve in opposite directions. Endochrome very pale. Vesicles wanting according to Meneghini; but Mr. Jenner informs me that they are present, although indistinct. The empty frond is colourless.

Mr. Jenner has gathered conjugated specimens at Rackham Bogs. The sporangium, which is quadrate, is large compared with the size of the plant. The empty segments of the frond remain attached to the angles of the sporangium.

Closterium Cornu differs from *C. acutum* in its obtuse ends.

Length of frond $\frac{1}{140}$ of an inch; breadth $\frac{1}{3709}$.

Length of β. $\frac{1}{226}$; breadth $\frac{1}{2142}$; length of sporangium $\frac{1}{816}$ to $\frac{1}{680}$; breadth $\frac{1}{1133}$ to $\frac{1}{906}$.

Tab. XXX. fig. 6. *a, b.* fronds of β. with endochrome; *c, d.* front views of sporangia; *e.* side view of sporangium; *f, g.* fronds of α.

22. **C. acutum** (Lyngbye); frond curved, gradually tapering at each extremity; end more or less acute; empty frond colourless.

α. Frond six to twelve times longer than broad; ends subacute.

Echinella acuta, Lyngbye, *Tent. Hydrophytologiæ Danicæ*, p. 209. t. 69. G. (1819).
Frustulia acuta, Kützing, *Synopsis Diatomearum in Linnæa* 1833, p. 537.
Closterium acutum, Brébisson, *in lit. cum icone* (1845).
Stauroceras acutum, Brébisson, *in lit.* (1846).

β. Frond ten to twenty times longer than broad, tapering at each extremity into a very fine point.

Frustulia subulata, Kützing, *Synop. Diatom.* p. 538. f. 3 (1833).
Stauroceras subulatum, Brébisson, *in lit. cum icone* (1846).

Dolgelley and Penzance, *J. R.* Rochdale, *Mr. Coates.* Brightling near Battle, Sussex, *Mr. Jenner.*

Germany, *Kützing.* Falaise, *Brébisson.*

Frond very minute and slender, many times longer than broad, narrow-lanceolate or acicular, slightly curved, gradually tapering; extremities hyaline, and more or less acute at the apex. Vesicles obscure; fillets none; endochrome very pale, and not extending to the extremities. The moving granules are free, and more or less remote from the ends.

The empty frond is quite colourless and free from markings.

Sporangia are not uncommon. I have frequently gathered them at Dolgelley and Penzance, and M. de Brébisson has sent me drawings of French specimens. The sporangium is cruciform in the front and oval in the side view. It is inclosed in a cell, similar to itself in form, but extending beyond it at the angles into the adjoining portion of the respective segments. In their empty state the segments have so much the colour and appearance of the water in which they are immersed that it is difficult to trace them, and the portions of the cell which project into the segments may be mistaken for them. The same circumstance renders it difficult to determine the exact form of the attenuated ends.

By Brébisson *Closterium subulatum* is regarded as a distinct species; and no doubt extreme states differ considerably, but I am unable to detect any distinctive character on which I can rely.

Closterium acutum differs from *C. Cornu* in its tapering frond and acute ends; its fronds are also far more closely united with the sporangium.

Length of frond $\frac{1}{177}$ of an inch; greatest breadth $\frac{1}{2550}$; length of sporangium $\frac{1}{1131}$ to $\frac{1}{645}$; breadth $\frac{1}{2109}$.

Tab. XXX. fig. 5. *a, b.* fronds of α; *c.* frond of β; *d.* conjugated fronds from a drawing by Brébisson; *e.* front view of sporangium; *f.* side view of sporangium from a drawing by Brébisson. Tab. XXXIV. fig. 5. sporangium.

17. SPIROTÆNIA, *Bréb.*

Frond simple, elongated, cylindrical or fusiform, straight, entire, not constricted at the middle; *ends* rounded; *endochrome* spiral.

The fronds are simple, straight, cylindrical, slightly attenuated at the extremities; they are quite entire and not constricted at the middle. The endochrome is spiral, as in *Zygnema*, and, except in fronds about to divide, there is no interruption at the centre.

The spiral arrangement of the endochrome will distinguish *Spirotænia* from every other genus in the family. It also differs from *Tetmemorus* in the entire extremities, and from that genus as well as from *Docidium* in the absence of a constriction at the centre. It differs likewise from *Closterium* by its straight fronds and the absence of a globule near the extremities.

Spirotænia affords another example of resemblance to the Conjugatæ. The endochrome is so exactly like that of *Zygnema*, that, until we saw M. de Brébisson's drawing, both Mr. Jenner and myself had regarded the *S. condensata* as the young state of a *Zygnema*. It however does not form a filament, the cells dividing in the manner of the Desmidieæ; but the division is oblique, as in *Scenedesmus*; thus affording a character by which it may be distinguished from *Penium*, even should the spiral form of the endochrome be absent.

In *Spirotænia* the complete division of the cell is prior to the division of the gelatinous covering, which thus retains the two newly parted cells together for some time longer; a fact which convincingly proves that this genus belongs to the Desmidieæ.

1. **S. condensata** (Bréb.); endochrome a single, broad, closely spiral band.

Spirotænia condensata, Brébisson, *in lit. cum icone* (1846).

Common. Dolgelley and Penzance, *J. R.* Henfield, Sussex, and bogs at Fisher's Castle, Tunbridge Wells, *Mr. Jenner.* Ayrshire, *Rev. D. Landsborough.* Aberdeenshire, *Dr. Dickie* and *Mr. P. Grant.* Ambleside, *Mr. Sidebotham.* Near Bristol, *Mr. Thwaites.*

Falaise, *Brébisson.*

Frond bright yellow-green, cylindrical with rounded ends, or fusiform, five to ten times longer than broad, enclosed in an evident mucous covering. The endochrome consists of a single broad spiral band. The revolutions are close, and from seven to twelve in number.

This is a beautiful plant; and in a recent state the spiral arrangement of the endochrome is very conspicuous, but it collapses shortly after being gathered.

Length of frond $\frac{1}{208}$ of an inch; breadth $\frac{1}{1048}$.

Tab. XXXIV. fig. 1. *a, b.* mature frond; *c.* fronds after division.

2. **S. obscura** ——; endochrome at first in several slender spiral threads, afterwards uniform.

Dolgelley and Penzance, *J. R.* Bogs at Fisher's Castle, Tunbridge Wells, *Mr. Jenner.*

Frond dark green, cylindrical or fusiform, extremities attenuated, five to eight times longer than broad. When young the endochrome is distinctly spiral, and the frond resembles a joint of a many-spired species of *Zygnema*. The threads or fillets are several in number, slender, and each of them makes but one or two revolutions in the length of the frond. Generally however the endochrome forms a uniform dark green mass without any appearance of spires. In this state, a small hyaline space left at each extremity will frequently be found to contain a small granule.

I long regarded this species as probably a variety of *Spirotænia condensata*, nor was it until I met with a young specimen that I was fully satisfied of its distinctness. Even specimens which do not exhibit the spires may be recognized by the darker colour of their endochrome and more tapering extremities.

Length of frond from $\frac{1}{247}$ to $\frac{1}{226}$ of an inch; greatest breadth from $\frac{1}{1020}$ to $\frac{1}{907}$.

Tab. XXXIV. fig. 2. *a, b, c.* single fronds; *d, e.* fronds after division.

*** *Cells fasciculated.*

18. ANKISTRODESMUS, *Corda.*

Cells elongated, attenuated, entire, aggregated in faggot-like bundles.

The cells are fusiform or crescent-shaped, have no constriction at their centre, and are precisely like those of *Closterium* except in their aggregation, which is the only distinction between the two genera.

1. *A. falcatus* (Corda); bundles of numerous crescent-shaped very slender cells.

Micrasterias falcata, Corda, *Almanach de Carlsbad* 1835, p. 121. t. 2. f. 29.
Closterium gregarium, Meneghini, *Consp. Alg. Eugan.* p. 17 (1837).
Xanthidium? difforme, Ehr. *Infus.* p. 147. t. 10. f. 26 (1838)?
Closterium falcatum, Meneghini, *Synop. Desmid. in Linnæa* 1840, p. 233.
Rhaphidium fasciculatum, Kützing, *Phycologia Germanica*, p. 144 (1845).
Ankistrodesmus gregarius, Brébisson, *in lit. cum specimine* (1846).

Dolgelley and Penzance, *J. R.* Sussex, *Mr. Jenner.* Ambleside, *Mr. Sidebotham.*

Italy, *Meneghini.* Germany, *Ehrenberg, Kützing.* Falaise, *Brébisson.*

Cells very minute and slender, crescent-shaped, fasciculated in irregular bundles composed of numerous individuals, which, as the convexity of each is turned inwards, diverge at each extremity of the bundle.

This plant is not unlike an early state of a *Closterium*, and as such I formerly considered it; but Corda, Kützing and Brébisson have pronounced it to belong to a distinct genus.

The bundles vary greatly in their compactness, as well as in the number of their cells.

Not having seen Corda's description and figure of *Ankistrodesmus convolutus*, I am unable to determine whether the present species differs from it.

Length of cell $\frac{1}{549}$ of an inch; greatest breadth $\frac{1}{7353}$.

Tab. XXXIV. fig. 3. *a, b, c.* aggregated cells; *d.* a single cell.

**** *Frond composed of few cells, definite in number, and not forming a filament.*

19. PEDIASTRUM, *Meyen.*

Frond plane, circular, composed of several cells, which form by their union a flattened star, and are generally arranged either in a single circle or in two or more concentric ones; marginal cells bipartite.

Frond minute, composed of four or more cells united together in the figure of a flattened star; when these are only four in number they are not arranged in a circle and the star is somewhat angular; in most species, however, they form either a single circle or two or more concentric ones, and one or two cells usually occupy the centre. The cells are combined into a frond by a mucous matrix, which is generally colourless and constitutes hyaline interstices. Occasionally some are ruptured, but in this case their endochrome alone escapes, and the others are not affected. In all the species the free margin of the outer cells is bipartite, a character which I consider important, and in fact a modification of the form observed in *Cosmarium* and other genera with constricted fronds. The cell in *Pediastrum*, however, is different from theirs, since its division into two segments exists only on one side, and in regard to the inner cells is at most a slight concavity of the external margin.

The flat, star-like fronds of *Pediastrum* are so characteristic, that there is no risk of mistaking it for any other genus, *Crucigenia* perhaps excepted; that however differs by having entire, quadrangular cells.

It is far more difficult to distinguish its species. Ehrenberg relies chiefly on the number of the circles, but this character, as Meneghini and Professor Bailey have observed, cannot always be depended on: the latter says, "There appears to me to be much confusion in the specific characters, arising from the circumstance that the number of corpuscles in the different rows has been made a character of specific importance. From what I have seen of the species, I am satisfied that the number of corpuscles in a star is liable to great variation in the same species."

I have myself noticed that, in one species at least, the number of circles varies from one to three; and the same observation has been made by Mr. Jenner. He has also remarked that the number of cells in the inner circle, on which I was inclined to place greater reliance, is subject to variation. Some species have one cell in the centre and others two; but the central cells are said to be sometimes deficient, a variation which I have not observed.

All the above characters, it must be allowed, are more or less uncertain; still upon them we must for the most part depend in discriminating between nearly allied species.

Meneghini adopts as a specific character the number and position of hyaline vesicles in each cell; but I am afraid that both of these are also variable.

The synonyms in this genus are so confused that I quote them with much hesitation. Ehrenberg, having relied almost entirely on the number of the circles and the cells in each, has neglected the form of the cells, and consequently his species are intermixed, states of some being referred to others. Mr. Hassall indeed has constituted several new species; but as he has attempted this from an examination merely of Ehrenberg's figures, and not of the plants themselves, his opinions are entitled to less weight than they would have been if founded on his own observation.

Professor Bailey has suggested, in his paper on the American Desmidieæ, that the form of the cells might afford the most certain character. Kützing, who has described nine species in his 'Phycologia Germanica,' seems to have arrived at the same conclusion. I have ventured in some instances to differ from Kützing's views; yet I feel persuaded that his definitions of the species in *Pediastrum* are truer to nature than those of Ehrenberg.

I have not seen the cells during the process of division; but I am informed by M. de Brébisson that it takes place at the notch, in the same manner as in other Desmidieæ: hence the cells in each circle are connected at their ends, like those of the filamentous genera. I do not however understand in what manner the additional circles are formed, nor why the numbers in each circle are so constant.

The segments or lobes are in some species more or less emarginate, and in others entire; as this character seems to be constant, I have taken advantage of it to divide the genus into two sections.

* *Lobes of the outer cells emarginate, or truncate.*

1. *P. Tetras* (Ehr.); cells four, separated by colourless interstices which form a cross; lobes truncato-emarginate.

Micrasterias Tetras, Ehr. *Infus.* p. 155. t. 11. f. 1 (1838). Bailey, *Amer. Journ. of Science and Arts*, v. 41. p. 293. t. 1. f. 19.
Pediastrum Tetras, Ralfs, *Annals of Nat. Hist.* v. 14. p. 469. t. 12. f. 4 (1844); *Trans. of Bot. Soc. of Edinburgh*, v. 2. p. 157. t. 17. Hassall, *Brit. Freshwater Alg.* p. 388.

Barmouth; Dolgelley and Penzance, *J. R.* Beckley Furnace near Battle, Sussex, *Mr. Jenner.* Ambleside, *Mr. Sidebotham.* Near Aberdeen, *Mr. P. Grant.*

Germany, *Ehrenberg.* Maine to Virginia, *Bailey.*

Frond extremely minute, composed of four cells, which constitute a star-like figure; the cells are somewhat triangular, and connected by a hyaline matrix, which forms colourless interstices in the figure of a cross; the free margin is bilobed; the lobes terminate in a broad shallow notch, which has acute corners. The colouring matter is pale green.

Meneghini makes *Pediastrum Tetras* a synonym of *P. Heptactis*; but I have not seen any intermediate forms, and in fact the frond of this species is sometimes as large as that of *P. Heptactis*, although the latter contains twice as many cells.

Length of cell $\frac{1}{2941}$ of an inch; breadth $\frac{1}{2272}$.

Tab. XXXI. fig. 1. *a, b.* fronds with endochrome; *c.* distorted frond; *d.* empty frond.

2. *P. Heptactis* (Ehr.); frond constituted of seven cells disposed in a circle round another in its centre; cells bipartite with emarginate lobes.

Micrasterias heptactis, Ehr. *Abh. d. Berl. Ak.* (1833), p. 300; *Infus.* p. 156. t. 11. f. 4. *b, c, d.*

Euastrum hexagonum, Corda, *Almanach de Carlsbad* 1835, p. 122. f. 31.

Pediastrum heptactis, Meneghini, *Synop. Desmid. in Linnæa* 1840, p. 211. Ralfs, *Annals of Nat. Hist.* v. 14. p. 469. t. 12. f. 5; *Trans. of Bot. Soc. of Edinburgh,* v. 2. p. 157. t. 17.

Pediastrum simplex, Hassall, *Brit. Freshwater Algæ,* p. 388 (1845); not of other authors.

Barmouth and Penzance, *J. R.* Beckley Furnace, Sussex, and near Southampton, *Mr. Jenner.* Rochdale, *Mr. Coates.* Near Aberdeen, *Mr. P. Grant.*

Germany, *Ehrenberg* and *Corda.* Mexico, *Bailey.*

Frond very minute, composed of eight cells, one in the centre and seven in a circle round it; the cells connected by a gelatinous matrix, which forms colourless interstices. The seven external cells are bipartite, their lobes ending in a broad shallow notch with acute corners; in other respects the cells are variable; the figure of the central one especially differs in almost every specimen, being sometimes angular and entire, and sometimes deeply divided with rounded segments. The colouring matter is very pale.

Length of cell $\frac{1}{2906}$ of an inch; breadth $\frac{1}{2500}$.

Tab. XXXI. fig. 2. *a, b.* fronds with endochrome; *c.* distorted frond; *d.* empty frond.

3. *P. biradiatum* (Meyen); outer circle generally of eleven bipartite cells, with bifid or emarginate lobes.

β. Lobes of outer cells truncato-emarginate.

Pediastrum biradiatum, Meyen, *Nov. Act. Acad. Nat. Cur.* v. 14. f. 21, 22 (1828). Meneghini, *Synop. Desmid. in Linnæa* 1840, p. 211.

Micrasterias Rotula, Ehr. *Infus.* p. 158. t. 11. f. 7 (1838).

Pediastrum Rotula, Kützing, *Phycologia Germanica,* p. 143 (1845).

α. Penzance, *J. R.* Ambleside, *Mr. Sidebotham.* Rochdale, *Mr. Coates.*

β. Beckley Furnace, Sussex, and near Southampton, *Mr. Jenner.*

Germany, *Meyen, Ehrenberg, Kützing.* New Jersey, *Bailey.*

Frond minute, generally composed of 16 or 17 cells arranged in two concentric circles, with a solitary cell in the centre; this central cell is said to be often wanting. I have never found it in British specimens, nor indeed even a vacancy in its place. The five angular cells forming the inner circle are often quadrilateral, and the exterior of each has a linear notch. The outer circle consists of eleven cells, which are either quadrilateral or somewhat cuneate;

the free margin is deeply bipartite, and the lobes are again divided or truncato-emarginate. The endochrome is bluish-green, with minute scattered granules. Interstices between the cells hyaline.

Kützing unites this and the two preceding species under Ehrenberg's name, *Rotula*; Meneghini keeps it distinct. Typal specimens of this species differ from *P. Tetras* and *P. Heptactis*, not only in the number of the cells, but also in the deeply divided and tapering lobes of the outer cells. Specimens however of the variety β, gathered in Sussex by Mr. Jenner, render it doubtful whether Kützing's view is not the correct one; for whilst these specimens have the same number and disposition of cells as *P. biradiatum*, the lobes of their marginal cells are merely slightly emarginate like those of *P. Heptactis*.

Length of cell in the outer circle $\frac{1}{2000}$ of an inch; breadth $\frac{1}{1754}$.

Length of β. $\frac{1}{2550}$; breadth $\frac{1}{2040}$.

Tab. XXXI. fig. 3. *a*. frond with endochrome; *b*. empty frond: fig. 4. *a*, *b*. fronds of β. with endochrome; *c*. empty frond.

** *Lobes of the segments entire.*

† *Connecting substance coloured.*

4. *P. Selenæa* (Kütz.); cells crescent-shaped, arranged in one or more circles around one or two central ones; connecting medium coloured.

Selenæa orbicularis, Nitzsch, Kützing, *Synopsis Diatom. in Linnæa* 1833, p. 604.
Micrasterias Selenæa, Kützing, *Synop. Diatom.* p. 604. f. 92 (1833).
Micrasterias Boryana, Ehrenberg, *Infusor.* t. 11. f. 5. *e* (1838).
Pediastrum Napoleonis, Ralfs, *Annals of Nat. Hist.* v. 14. p. 470. t. 12. f. 6 (1844); *Trans. of Bot. Soc. of Edinburgh*, v. 2. p. 158. t. 17; not of Meneghini.
Pediastrum Selenæa, Kützing, *Phycologia Germanica*, p. 143? (1845).
Pediastrum elegans, Hassall, *British Freshwater Algæ*, p. 389 (1845).
Pediastrum lunare, Hassall, *British Freshwater Algæ*, t. 92. f. 3 (1845).

Barmouth, North Wales, *J. R.*

Germany, *Kützing, Ehrenberg.*

Pediastrum Selenæa, like the other species of this genus, varies in the number of cells. Kützing's figures represent three states, having respectively one, two, and three circles, round a central cell, all the cells being lunate; the number of cells consisting of 5 in the first circle, 10 in the second, and 15 in the third.

The following description applies to British specimens only.

Frond minute, with two angular cells in the centre, and six crescent-shaped ones arranged in a circle round them. The cells are not approximate, or separated by hyaline interstices, as in other species; but are connected by a coloured portion, which indeed may be said to constitute the frond, the cells being imbedded in it. It will be seen that the British specimens differ from those figured by Kützing by having two cells in the centre instead of one, and six in the circle instead of five. I have never seen specimens with more than one circle. Notwithstanding these differences, I believe that our plant is correctly referred to this species. It agrees both in the crescent-shaped cells and in the coloured connecting medium; characters which distinguish it from all others.

Tab. XXXI. fig. 5. frond with endochrome.

†† *Interstices of the frond hyaline.*

5. **P. simplex** (Meyen); frond composed of four cells with or without one or two central ones; lobes of marginal cells ovate, tapering to a point.

    α. Marginal cells truncate at the base and forming a circle, its centre vacant or occupied by one or two cells.

Pediastrum simplex, Meyen, *Nov. Act. Acad. Nat. Cur.* v. 14. t. 42. f. 1–3 (1828). Kützing, *Phycologia Germanica*, p. 142.
Micrasterias simplex, Kützing, *Synopsis Diatom. in Linnæa* 1833, p. 602.
Micrasterias Coronula, Ehrenberg, *Infusor.* p. 156. t. 11. f. 2 (1838).
Pediastrum Napoleonis, Hassall, *Brit. Freshwater Algæ*, t. 92. f. 10, 11 (1845).

    β. *cruciatum*; cells angular at the base, connected in a cruciform manner without a central space.

Pediastrum cruciatum, Kützing, *Phycologia Germanica*, p. 142 (1845); *in lit. cum icone.*

    Near Aberdeen, *Mr. P. Grant.*
    Germany, *Kützing, Ehrenberg.*

In all the specimens that I have seen of this species, the number of cells was invariably four; but, according to Ehrenberg and other authors, it occasionally has one or two other cells in the centre. I have observed two forms: in one of them the cells were truncated at the base and the frond had a central square vacancy; in the other the bases of the cells extended into angles, and, as these exactly fitted each other, the interstices or joinings formed a cross. The latter state I believe to be the *Pediastrum cruciatum* of Kützing, although the figures which he sent me represent smaller fronds.

At the base of each cell in β. there is generally a large vesicle, which however is sometimes absent. At first sight this variety appears to be distinct, but I have seen intermediate specimens, and, in all, the lobes of the cells are ovate at the base and taper into a cuspidate point.

Length of cell including lobes $\frac{1}{1020}$ of an inch; breadth $\frac{1}{1632}$; length of β. $\frac{1}{1666}$; breadth from $\frac{1}{3448}$ to $\frac{1}{2040}$.

Tab. XXXIV. fig. 15. *a.* frond with endochrome; *b.* empty frond; *c.* frond of β; *d. P. cruciatum* from a drawing by Kützing.

6. **P. pertusum** (Kütz.); inner cells leaving hyaline intervals resembling foramina; outer cells rectangular, notch triangular, as broad as the cell.

Micrasterias Boryana, Ehr. *Infusor.* t. 11. f. 5. *b, d, i* (1838).

Micrasterias tricyclia, Ehr. *Infusor.* t. 11. f. 8. *a, b* (1838).
Pediastrum pertusum, Kützing, *Phycologia Germanica*, p. 143 (1845).
Pediastrum tricyclium, Hassall, *Brit. Freshwater Algæ*, t. 92. f. 1 (1845).

Aberdeen, *Mr. P. Grant.* Storrington, Sussex, *Mr. Jenner.*
Germany, *Ehrenberg.*

Frond consisting of from one to three circles of cells arranged round one or two central ones. The number of cells, as in *Pediastrum Selenæa* and *P. Boryanum*, generally consists of 5 in the first circle, 10 in the second, and 15 in the third; but I have seen fronds with four cells forming a circle about two central ones. Ehrenberg's figures represent all the cells as alike in form, and this was the case in specimens sent me by Mr. P. Grant; but Mr. Jenner finds the inner cells quadrate, and only the marginal ones notched. In all the varieties, however, the inner cells are so placed as to leave hyaline spaces which have the appearance of perforations. The marginal cells have, on their free side, a wide triangular notch, about equal to the fourth part of a square; were this filled it would make the cell quadrate, consequently the sides and base of the cell form rectangular triangles, and the terminal lobes taper to an acute point.

Length of cell in outer circle $\frac{1}{3266}$ of an inch; breadth $\frac{1}{3268}$.

Tab. XXXI. fig. 6. *a, b.* fronds with endochrome.

7. **P. granulatum** (Kütz.); cells granulated; lobes of marginal cells tapering.

Pediastrum granulatum, Kützing, *Phycologia Germanica*, p. 143 (1845); *in lit. cum icone.*

Weston Bogs near Southampton, *Mr. Jenner.*
Germany, *Kützing.*

Frond composed of six cells circularly arranged around two subquadrate central ones; the marginal cells have lobes gradually tapering into short points, and all are rough with minute granules.

I have seen no specimens of this species, but Mr. Jenner's drawings exactly agree with one sent me by Professor Kützing.

Length of cell in outer circle $\frac{1}{2000}$ of an inch; breadth $\frac{1}{1851}$.

Tab. XXXI. fig. 8. *a.* frond with endochrome; *b.* empty frond.

8. **P. Napoleonis** (Turpin); six angular cells forming a circle round two central ones; lobes of marginal cells cuspidate; notch wide.

Helierella Napoleonis, Turpin, *Dict. des Sc. Nat. par Levr. Atl. Veg. ves.* fig. 20 (1820); *Mém. du Mus.* v. 16. t. 13. f. 21.
Micrasterias Napoleonis, Kützing, *Synop. Diatom. in Linnæa* 1833, p. 602. Ehrenberg, *Infus.* p. 156. t. 11. f. 3.
Pediastrum Napoleonis, Meneghini, *Synop. Desmid. in Linnæa* 1840, p. 212.
Pediastrum hexactis, Hassall, *Brit. Freshwater Algæ*, t. 92. f. 5 (1845).
Pediastrum excavatum, Hassall, *l. c.* t. 92. f. 6 (1845).

Ashdown Forest, Sussex, *Mr. Jenner.* Swansea, *J. R.* Near Aberdeen, *Mr. P. Grant.*

Germany, *Kützing, Ehrenberg.*

I greatly doubt whether this species is distinct from *Pediastrum Boryanum*. Its claim to rank as a species depends on the number and arrangement of its cells. The cells themselves are variable in form. The notch of the marginal ones is rounded, and usually wider than in the allied species.

Length of cell in outer circle from $\frac{1}{1569}$ to $\frac{1}{1483}$ of an inch; breadth from $\frac{1}{1813}$ to $\frac{1}{1088}$.

Tab. XXXI. fig. 7. *a, b, c.* fronds with endochrome; *d, e.* empty fronds.

9. *P. angulosum* (Ehr.); marginal cells with angular lobes which are not extended into rays; interstices hyaline.

Micrasterias angulosa, Ehrenberg, *Abh. der Berlin Ak.* p. 301 (1833); *Infusor.* p. 156. t. 11. f. 6. *a:*
Pediastrum angulosum, Meneghini, *Synopsis Desmid. in Linnæa* 1840, p. 211. Hassall, *British Freshwater Algæ*, p. 391.
Pediastrum Boryanum, Ralfs, *Annals of Nat. Hist.* v. 14. t. 12. f. 7 (upper figure) (1844); *Trans. of Bot. Soc. of Edinburgh*, v. 2. t. 17.

Dolgelley, *J. R.*   Ambleside, *Mr. Sidebotham.*   Aberdeen, *Mr. P. Grant.* Beckley Furnace near Battle, Sussex, *Mr. Jenner.*

Germany, *Ehrenberg.*

Frond minute, consisting of one or more circles of cells, with a single cell in the centre; the number of cells are generally five in the inner circle, ten in the second, and fifteen in the third; the inner cells are angular, and similar to those of *Pediastrum Boryanum* and *P. vagum.* The marginal cells are less deeply notched than in those species, and the lobes are merely angular, not in the least extended into processes or rays. The angle is produced by the sloping of the outer margin, which frequently might be termed obliquely truncate, or almost emarginate.

Length of cell in outer circle $\frac{1}{2732}$ of an inch; breadth $\frac{1}{1942}$.

Tab. XXXI. fig. 11. *a.* frond with endochrome; *b,* empty frond.

10. *P. Boryanum* (Turpin); cells arranged in one or more circles round one or two central ones; marginal cells gradually tapering into two long subulate points; notch narrow.

Helierella Boryana, Turpin, *Dict. des Sc. Nat. par Levr. Atl. Veg. ves.* f. 22 (1820); *Mém. du Mus.* v. 16. t. 13. f. 22.
Micrasterias Boryi, Kützing, *Synop. Diatom. in Linnæa* 1833, p. 603.
Micrasterias Boryana, Ehrenberg, *Abh. der Berlin Ak.* p. 300 (1833); *Infus.* p. 157. t. 11. f. 5. *a, c, g, h.*
Pediastrum Boryanum, Meneghini, *Synop. Desmid. in Linnæa* 1840, p. 210. Ralfs, *Annals of Nat. Hist.* v. 14. p. 470. t. 12. f. 7 (lower figure); *Trans. of Bot. Soc. of Edinburgh*, v. 2. p. 158. t. 17. Kützing, *Phyc. Germ.* p. 143. Jenner, *Fl. of Tunbridge Wells*, p. 198. Hassall, *Brit. Freshwater Algæ*, p. 389.

Dolgelley, Barmouth, and Penzance, *J. R.*   Cheshunt, *Mr. Hassall.* Beckley Furnace near Battle, Storrington, Tunbridge Wells, &c., Sussex; Dorking, Surrey; and near Southampton, *Mr. Jenner.* Manchester, *Mr. Gray* and *Mr. Williamson.*

Germany, *Kützing, Ehrenberg.* Falaise, *Brébisson.* Maine to Mexico, *Bailey.*

Cells generally arranged in one to three circles round a central cell. The number of cells consists of five in the first circle, ten in the second, and fifteen in the third; but these numbers are not invariable; sometimes there are two central cells, and still more frequently the numbers in the circles differ from those just stated. The inner cells are five- or six-angled, broader than long, with a slight concavity on the outer margin; the marginal cells are biradiate, the rays subulate, the notch narrow.

Pediastrum Boryanum differs from *P. vagum* in the gradually tapering acute lobes of the outer cells; from *P. angulosum* it may be known by the lobes, which are elongated into rays, and not merely angular, as in that species.

Length of cell in outer circle from $\frac{1}{2083}$ to $\frac{1}{1633}$ of an inch; breadth from $\frac{1}{2732}$ to $\frac{1}{2222}$.

Tab. XXXI. fig. 9. *a.* frond with endochrome; *b.* empty frond.

11. **P. ellipticum** (Ehr.); cells variable in number and arrangement; lobes of marginal ones suddenly contracted into short, cylindrical, obtuse processes.

β. processes of the lobes truncato-emarginate.

Micrasterias elliptica, Ehrenberg, *Abh. der Berlin Ak.* p. 302 (1833); *Infus.* p. 158. t. 11. f. 9.
Micrasterias, No. 3, Bailey, *American Journal of Science and Arts*, v. 41. p. 293. t. 1. f. 21 (1841).
Pediastrum Boryanum, Ralfs, *Annals of Nat. Hist.* v. 14. t. 12. f. 8 (1844); *Trans. of Bot. Soc. of Edinburgh*, v. 2. t. 17.
Pediastrum vagum, Kützing, *Phycologia Germanica*, p. 143 (1845); *in lit. cum icone.*
Pediastrum constrictum, Hassall, *Brit. Freshwater Algæ*, p. 391 (1845).
Pediastrum ellipticum, Hassall, *Brit. Freshwater Algæ*, t. 92. f. 2 (1845).

Dolgelley, *J. R.* Aberdeenshire and Banffshire, *Mr. P. Grant.* Near Bristol, *Mr. Thwaites.* Beckley Furnace near Battle, Sussex; Dorking, Surrey; and near Southampton, *Mr. Jenner.*

Germany, *Ehrenberg, Kützing.* Maine to Virginia, *Bailey.*

Frond larger than that of any other species, often very irregular in form; cells numerous, but varying much in number; the arrangement of the inner ones, especially in the larger specimens, is more or less irregular, and not in distinct circles, like those in the preceding species. The notch of the marginal cells is narrow, the lobes terminated by a cylindrical process, which appears to me rounded at its apex; but Mr. Jenner usually finds two minute teeth at the end.

Length of cell in outer circle from $\frac{1}{1754}$ to $\frac{1}{906}$ of an inch; breadth from $\frac{1}{1515}$ to $\frac{1}{1020}$.

Tab. XXXI. fig. 10. *a, b, c.* fronds with endochrome; *d.* empty frond.

20. SCENEDESMUS, *Meyen*.

Frond composed of two to ten fusiform or oblong cells, arranged side by side in a single row, but after division in two alternating rows; division oblique.

Cells always entire, and mostly fusiform or oblong, but in some species the outer ones are lunulate. There is no constriction nor suture at the middle, and the endochrome is not divided into two portions by a transverse band. The cells are few in number, varying from two to ten, and united into a frond by a hyaline matrix. Their division is oblique, and not transverse, as in most genera of the Desmidieæ. As all the cells divide simultaneously, the frond when dividing consists of two rows, which are retained in connection by the matrix for some time after the complete division of the cells. From the oblique manner of division, the cells of one row alternate with those of the other.

The dividing frond is so unlike one having only a single row of cells, that, in common with Meneghini and other naturalists, I formerly considered these states as different species; and it was only on seeing a frond of *Scenedesmus obtusus*, in which the division of the cells was still incomplete, that I discovered my error. The fact however was previously known to Brébisson, and he has since sent me drawings illustrating the process.

The endochrome is in general very pale, and the starch granules are inconspicuous.

Scenedesmus differs from the preceding genera in the very different form of its cells, but *Pediastrum* supplies a connecting link between them. As in that genus, the frond in *Scenedesmus* is composed of several cells, but these are differently arranged; and the division into two segments, which, although modified, is still met with in the outer cells of *Pediastrum*, is entirely absent in *Scenedesmus*. In the oblique manner in which its cells divide, it agrees with *Spirotænia*, which however has a different arrangement of the endochrome, and a frond consisting merely of a single cell.

1. *S. quadricauda* (Turp.); cells generally four, oblong, rounded at their ends, disposed in a single row; each extremity of the two external ones usually terminated by a bristle.

Achnanthes quadricauda, Turpin, *Dict. des Sc. Nat. par Levr. Atl. Veg. ves.* f. 8 (1820); *Mém. du Mus.* f. 6.
Scenedesmus magnus et S. longus, Meyen, *Nov. Act. Acad. Nat. Cur.* v. 14. t. 43. f. 26–29 (1828). Kützing, *Synop. Diatom. in Linnæa* 1833, p. 606.
Scenedesmus quadricaudatus α. *cornutus*, Ehr. *Abh. d. Berl. Ak.* 1833, p. 309. Ralfs, *Annals of Nat. Hist.* v. 15. p. 402. t. 12. f. 4; *Trans. of Bot. Soc. of Edinburgh*, v. 2. p. 159. t. 15. Hassall, *Brit. Freshwater Algæ*, p. 393.
Scenedesmus quadricauda, Brébisson, *Alg. Fal.* p. 66 (1835). Menegh. *Synop. Desmid. in Linnæa* 1840, p. 206.
Scenedesmus caudatus, Corda, *Alm. de Carlsb.* 1835, p. 123. t. 4. f. 50. Kützing, *Phy. Germ.* p. 139.
Arthrodesmus quadricaudatus, Ehr. *Infus.* p. 150. t. 10. f. 16 (1838). Bailey, *Amer. Journ. of Science and Arts*, v. 41. p. 292. t. 1. f. 17.

β. External cells with three bristles.

γ. *ecornis* (Ehr.); all the cells similar, and without bristles.

Achnanthes quadrijuga, Turpin, *Dict. des Sc. Nat.* f. 5 (1820).
Scenedesmus Leibleinii, Kütz. *Leib. Bot. Zeit.* 1830, f. 3; *Synop. Diatom.* p. 607. Menegh. *Synop. Desmid.* p. 207.
Scenedesmus quadricaudatus β. *ecornis*, Ehr. *Abh. d. Berl. Ak.* 1833, p. 309.

Common. Wales and Cornwall, *J. R.* Sussex; Kent; and near Southampton, *Mr. Jenner.* Bristol, *Mr. Thwaites* and *Mr. Broome.* Herts, *Mr. Hassall.* Ayrshire, *Rev. D. Landsborough.* Aberdeenshire and Banffshire, *Mr. P. Grant.* Northamptonshire, *Rev. M. J. Berkeley.* Kerry, *Mr. Andrews.* Manchester, *Mr. Williamson.* Ambleside, *Mr. Sidebotham.* Rochdale, *Mr. Coates.*

Germany, *Kützing, Ehrenberg*, &c. Falaise, *Brébisson.* Maine to Virginia, and in Mississippi River, *Bailey.*

Frond composed of from four to eight oblong cells, which are generally larger than those of any other species in the genus, about three times as long as broad, and rounded at their ends; the external cells are usually the most turgid, and their bristles are directed outwards.

The colouring matter is pale, with minute scattered granules.

In β. the cells are smaller, and the external ones, besides the usual terminal bristles, have another from the centre of the outer margin.

The variety γ. is described by some authors as a distinct species; but I agree with Ehrenberg in considering it a state of *S. quadricauda*, from which it differs only in having no bristles.

Length of cell $\frac{1}{1121}$ of an inch; breadth $\frac{1}{2631}$.

Tab. XXXI. fig. 12. *a, b, c, d.* fronds with endochrome; *e, f.* empty fronds; *g.* variety β; *h.* variety γ; *i.* dividing frond of γ.

2. *S. dimorphus* (Turp.); cells acute, four to eight, placed evenly in a single row; the inner cells fusiform, the outer externally lunate.

Achnanthes dimorpha, Turp. *Dict. des Sc. Nat. par Levr. Atl. Veg. ves.* f. 7 (1820).
Scenedesmus pectinatus, Meyen, *Nov. Act. Acad. Nat. Cur.* v. 14. p. 2. f. 34, 35 (1828). Kützing, *Synop. Diatom.* p. 608; *Phy. Germ.* p. 140. Brébisson *in lit.*
Scenedesmus dimorphus, Kützing, *Synop. Diatom. in Linnæa* 1833, p. 608. Meneghini, *Synop. Desmid. in Linnæa* 1840, p. 208. Ralfs, *Annals of Nat. Hist.* v. 15. p. 403. t. 12. f. 5; *Trans. of Bot. Soc. of Edinburgh*, v. 2. p. 160. t. 15. Hassall, *Brit. Freshwater Algæ*, p. 393.
Arthrodesmus pectinatus, Ehr. *Infus.* p. 151. t. 10. f. 17 (1838).

Dolgelley and Penzance, *J. R.* Bristol, *Mr. Thwaites.* Manchester, *Mr. Sidebotham* and *Mr. Williamson.* Aberdeen, *Mr. P. Grant.* Weston Bogs near Southampton, *Mr. Jenner.*

Germany, *Kützing, Ehrenberg, Corda,* &c. Falaise, *Brébisson.*

Frond very minute, consisting of four to eight cells placed evenly side by side in a single row; the inner cells straight, fusiform, attenuated, and acute at each end; the outer ones externally lunate. The endochrome is pale bluish-green.

Length of cell from $\frac{1}{1020}$ to $\frac{1}{900}$ of an inch; breadth $\frac{1}{8160}$.

Tab. XXXI. fig. 13. *a.* frond with endochrome; *b.* empty frond.

3. *S. acutus* (Meyen); cells two to eight, fusiform, acuminate, arranged in a single, irregularly alternating series.

Scenedesmus acutus, Meyen, *Nov. Act. Acad. Nat. Cur.* v. 14. t. 43. f. 32 (1828). Kützing, *Synop. Diatom. in Linnæa* 1833, p. 609. f. 96; *Phycologia Germanica*, p. 139. Meneghini, *Synop. Desmid. in Linnæa* 1840, p. 207. Ralfs, *Annals of Nat. Hist.* v. 15. p. 403. t. 12. f. 6; *Trans. of Bot. Soc. of Edinburgh*, v. 2. p. 160. t. 15. Hassall, *Brit. Freshwater Algæ*, p. 393.
Scenedesmus fusiformis, Meneghini, *Consp. Alg. Eugan.* p. 18 (1837); *Synop. Desmid. in Linnæa* 1840, p. 208.
Arthrodesmus acutus, Ehrenberg, *Infusor.* p. 150. t. 10. f. 19 (1838).

King's Cliffe, Northamptonshire, *Rev. M. J. Berkeley.* Bristol, *Mr. Thwaites.* Dolgelley, *J. R.* Weston Bogs near Southampton, *Mr. Jenner.*

Germany, *Ehrenberg, Kützing.* Falaise, *Brébisson.*

Cells fusiform, acute at each end, frequently more or less ventricose; after division forming a single series, but projecting alternately more or less on each margin. The two outer cells are frequently crescent-shaped.

When the cells are nearly uniform this species has some resemblance to *Scenedesmus dimorphus*; but in the latter the cells are more slender, never ventricose, and are arranged quite evenly side by side. It is more difficult to distinguish *S. acutus* from *S. obliquus*, and I am far from certain that Ehrenberg erred in uniting them. The principal distinction is that in *S. acutus* the cells form only a single series, which is nevertheless irregular on account of the alternate projection of the cells in opposite directions. In *S. obliquus*, on the other hand, the cells by division form two distinct rows, which, after separation, become two fronds.

Length of cell from $\frac{1}{1063}$ to $\frac{1}{1020}$ of an inch; breadth from $\frac{1}{6181}$ to $\frac{1}{5555}$. Tab. XXXI. fig. 14. perfect frond. Tab. XXXIV. fig. 16. distorted variety.

4. *S. obliquus* (Turpin); cells elliptico-fusiform, after division arranged in two distinct, generally oblique series, the outermost cell of each not in contact with any of those in the other series.

Achnanthes obliqua, Turpin, *Dict. des Sc. Nat. par Levr. Atl. Veg. ves.* f. 9 (1820); *Mém. du Mus.* v. 16. t. 13. f. 9.
Scenedesmus obliquus, Kützing, *Synop. Diatom. in Linnæa* 1833, p. 609. Meneghini, *Synop. Desmid. in Linnæa* 1840, p. 208. Berkeley, *English Botany*, t. 2933.
Scenedesmus triseriatus, Meneghini, *Conspect. Alg. Eugan.* p. 18 (1837); *Synop. Desmid. in Linnæa* 1840, p. 208. Ralfs, *Annals of Nat. Hist.* v. 15. p. 403. t. 12. f. 7; *Trans. of Bot. Soc. of Edinburgh*, v. 2. p. 161. t. 15.
Arthrodesmus acutus, Ehrenberg, *Infusor.* t. 10. f. 19. *b* (1838).

King's Cliffe, *Rev. M. J. Berkeley*. Near Bristol, *Mr. Thwaites*. Ayrshire, *Rev. D. Landsborough*. Near Manchester, *Mr. Sidebotham*. Dolgelley and Penzance, *J. R.* Near Aberdeen, *Mr. P. Grant* and *Dr. Dickie*. Storrington, Sussex; and Weston Bogs near Southampton, *Mr. Jenner*.

Germany, *Kützing, Ehrenberg*. Italy, *Meneghini*. Falaise, *Brébisson*.

Scenedesmus obliquus approaches very closely to *Scenedesmus acutus*, and the principal distinction depends on the different appearances presented by the dividing frond; and as this plant is most commonly seen in that state, I shall so describe it, merely premising that when separated, the frond consists of only one row. Mr. Berkeley has united *S. triseriatus* to *S. obliquus*, and I readily follow his example.

The cells are eight in number, arranged obliquely in two distinct series. Each series has three cells, which are fusiform, equal, somewhat ventricose in the middle, acute at the free extremity, and subacute or rounded at the inner one. As the cells are not placed evenly, but each projects beyond its neighbour, the rows are oblique; their position with respect to each other is such that the inner ends of two cells of the lower lie between those of the upper row, whilst that of the third is outside the end of the highest in the upper. Of the two remaining cells, which are lunate, one is placed beyond the outer cell in each series, and is therefore not in contact with any cell in the other. They however take the same direction as the other cells in the series to which they respectively belong.

The different species of *Scenedesmus* frequently make their appearance in clear water that is kept in glasses or bottles and exposed to the light. I have received specimens of this species collected in this manner from Mr. Landsborough and Mr. Berkeley, and have myself repeatedly noticed its appearance in bottles containing other Desmidieæ, and sometimes its rapid increase so as to outnumber its companions. I must however observe, that specimens thus obtained are more variable in their characters than such as are gathered in their native abodes, and that their cells are frequently more or less distorted.

Length of cell $\frac{1}{1666}$ of an inch; breadth $\frac{1}{6250}$.

Tab. XXXI. fig. 15. *a, b, c.* different states of frond.

5. **S. obtusus** (Meyen); cells three to eight, ovate or oblong, and arranged in one row, or, after division, alternately in two rows.

Achnanthes quadralterna et octalterna, Turp. *Dict. des Sc. Nat. par Levr. Atl. Veg. ves.* f. 7, 8 (1820).
Scenedesmus obtusus, Meyen, *Nov. Act. Acad. Nat. Cur.* v. 14. f. 31 (1828). Meneghini, *Synop. Desmid. in Linnæa* 1840, p. 208. Ralfs, *Annals of Nat. Hist.* v. 15. p. 404. t. 12. f. 8; *Trans. of Bot. Soc. of Edinburgh,* v. 2. p. 162. t. 15. Kützing, *Phy. Germ.* p. 139. Hassall, *Brit. Freshwater Algæ,* p. 394.
Scenedesmus quadralternus, Kütz. *Synop. Diatom. in Linnæa* 1833, p. 608. f. 94.
Scenedesmus octoalternus, Kützing, *l. c.* p. 609. f. 95.
Arthrodesmus acutus?, Bailey, *Amer. Journ. of Science and Arts,* v. 41. p. 292. t. 1. f. 18 (1841).

Common. Dolgelley and Penzance, *J. R.* Storrington and Beckley Furnace near Battle, Sussex; Reigate, Surrey; and near Southampton, *Mr. Jenner.* Bristol, *Mr. Thwaites.* Manchester, *Mr. Williamson.* Ambleside, *Mr. Sidebotham.* Aberdeenshire, *Mr. P. Grant.*

Germany, *Kützing.* Falaise, *Brébisson.* Maine to Virginia, and in the Mississippi River, *Bailey.*

Frond minute, composed of from three to eight ovate or ovato-oblong cells with rounded ends. The endochrome is very pale green.

This species is rarely met with in a simple state, but as the cells after division are still retained in connection by the mucous matrix, two rows are usually present, the broader ends of one row lying between the cells of the other. The hyaline matrix is frequently their only bond of union, and in this state they seem to connect the Desmidieæ with the Ulvaceæ through *Merismopedia.*

Length of cell from $\frac{1}{2331}$ to $\frac{1}{1961}$ of an inch; greatest breadth from $\frac{1}{4096}$ to $\frac{1}{3623}$.

Tab. XXXI. fig. 16. *a, b.* ordinary state of frond; *c.* frond just divided.

6. **S. duplex** (Kütz.); cells slender, fusiform, sigmoid, tapering at each end into a fine point; after division closely connected for about half their length.

Rhaphidium duplex, Kützing, *Phycologia Germanica,* p. 144 (1845).

Penzance, *J. R.*

Germany, *Kützing.*

Cells linear-lanceolate; extremities tapering to a fine point and curved in opposite directions. The cells after division remain closely united; frequently the frond consists of only a single pair of cells so connected, but sometimes of two or even three of these pairs, which however are remote from each other: in this case, as the connecting mucus is colourless, they look like distinct plants, and their relation can be detected only by moving the frond. If kept in water for a few days, the cells separate from each other.

Professor Kützing formed his genus *Rhaphidium* for the reception of the plant above described and a species of *Ankistrodesmus*; but I believe the proper position of the former is in this genus.

Tab. XXXIV. fig. 17. *a.* frond with dividing cells; *b.* cell with the division more advanced.

ANALYSIS OF THE GENERA.

1. Frond a single cell (its division being followed by separation), or cells fasciculated 2
   Frond composed of several cells 13

2. Segments, in the front view of the frond, lobed, incised, sinuated, or notched 3
   Segments, in the front view, undivided........................ 4

3. Frond fusiform; segments notched at the end, otherwise entire .... *Tetmemorus.*
   Frond oblong, compressed; segments sinuated .... *Euastrum.*
   Frond lenticular; lobes of segments incised or bidentate at the apex .... *Micrasterias.*

4. End view of frond angular or radiate 5
   End view of frond neither angular nor radiate 6

5. Processes geminate from each angle, one external to the other .... *Didymocladon.*
   Processes, if present, with their divisions on the same plane.. *Staurastrum.*

6. Cells short, rarely more than twice as long as broad 7
   Cells elongated (cylindrical, fusiform or lunately curved) 9

7. Frond with elongated, simple or branched spines 8
   Frond not spinous .... *Cosmarium.*

8. Segments with two spines .... *Arthrodesmus.*
   Segments with four or more spines .... *Xanthidium.*

9. Cells straight 10
   Cells curved... 12

10. Endochrome spiral .... *Spirotænia.*
    Endochrome not spiral................................... 11

11. Frond moderately elongated, if linear not constricted at the middle, and its segments not inflated at the base .... *Penium.*
    Frond much elongated, linear, generally much constricted at the middle, and its segments inflated at the base........ *Docidium.*

o

12. Cells not fasciculated .... *Closterium.*
    Cells fasciculated .... *Ankistrodesmus.*
13. Frond an elongated jointed filament 14
    Frond of few cells not forming a filament 18
14. Filament plane or triangular, with junction-glands or foramina between the joints 15
    Filament never plane; neither glands nor foramina between the joints 16
15. Joints with junction-glands; their margins incised or sinuated .... *Sphærozosma.*
    Joints with foramina between them; their margins bicrenate or entire .... *Aptogonum.*
16. Filament triangular or quadrangular .... *Desmidium.*
    Filament cylindrical or subcylindrical 17
17. Joints surrounded by a groove .... *Hyalotheca.*
    Joints with two opposite bicrenate projections .... *Didymoprium.*
18. Cells arranged in the form of a flattened star .... *Pediastrum.*
    Cells placed side by side in one or two rows .... *Scenedesmus.*

ANALYSIS OF THE SPECIES.

1. HYALOTHECA.

Fragile; joints bicrenate .... *dissiliens.*
Scarcely fragile; joints with a minute bifid projection. ...... *mucosa.*

2. DIDYMOPRIUM.

Joints broader than long; with angular teeth .... *Grevillii.*
Joints longer than broad; with rounded crenatures .... *Borreri.*

3. DESMIDIUM.

Filament triangular .... *Swartzii.*
Filament quadrangular .... *quadrangulatum.*

4. APTOGONUM.

.... *Desmidium.*

5. SPHÆROZOSMA.

Constriction of joints a linear notch on each margin .... *vertebratum.*
Constriction a broad sinus on each margin .... *excavatum.*

6. MICRASTERIAS.

1. Lobes of segments radiate, incised and cuneate 2
   Lobes horizontal, tapering and bidentate 12
2. Frond orbicular or oblong; lobes and subdivisions approximate 3
   Frond subelliptic; end lobe exserted 10
3. Frond oblong... *Jenneri.*
   Frond orbicular....................................... 4
4. End lobes broadest 5
   End lobes narrow...................................... 6
5. End margin convex .... *crenata.*
   End margin truncate or concave..................... *truncata.*
6. Principal divisions of the frond bordered with puncta-like
   granules .... *papillifera.*
   Surface of frond smooth 7
7. Ultimate subdivision of lateral lobes inflated and tapering .... *radiosa.*
   Ultimate subdivisions truncate............................ 8
8. Extremities of subdivisions neither spinous nor mucronate.. *denticulata.*
   Extremities of subdivisions either spinous or mucronate 9
9. Extremities of subdivisions abrupt with acute or spinous angles.. *rotata.*
   Extremities of subdivisions rounded and furnished with two
   mucros .... *fimbriata.*
10. Angles of end lobe bipartite; lateral lobes toothed......... *americana.*
    Angles of end lobe bidentate; lateral lobes deeply divided 11
11. Subdivisions of lobes long, narrow-linear .... *furcata.*
    Subdivisions of lobes short, scarcely linear.......... *Crux-Melitensis.*
12. End margin convex; lobes conical .... *oscitans.*
    End margin straight; lobes triangular .... *pinnatifida.*

7. EUASTRUM.

1. Terminal lobe distinct, cuneate, included in a notch formed
   by the lateral lobes 2
   Terminal lobe exserted or indistinct 4
2. Frond rough; end margin sinuated .... *verrucosum.*
   Frond smooth; terminal notch linear...................... 3
3. Segments 5-lobed .... *oblongum.*
   Segments 3-lobed .... *crassum.*
4. Segments with a distinct, narrow, terminal notch 5
   Segments with the end margin straight or slightly concave 15

5. Ends protuberant or pouting; angles usually terminated by
   a spine 6
   End straight or convex 7
6. End projections angular and tooth-like .... *rostratum.*
   End projections pouting and rounded .... *elegans.*
7. End lobe distinct, dilated, and connected to the basal portion by a distinct neck. 8
   End lobe obscure, not dilated 12
8. Transverse view with straight or concave sides; neck very long .... *insigne.*
   Transverse view with inflated or lobed middle 9
9. Segments distinctly 5-lobed .... *pinnatum.*
   Segments not 5-lobed 10
10. Neck of segments included between the process-like elongations of middle lobes .... *humerosum.*
    Neck exserted 11
11. Segments inflated at the base .... *ampullaceum.*
    Segments pyramidal, not inflated...................... *affine.*
12. End margin truncate, with acute angles................. *binale.*
    Ends rounded 13
13. Segments cuneiform, not lobed .... *cuneatum.*
    Segments either lobed or contracted upwards into a neck (not cuneiform) 14
14. Transverse view with one protuberance on each side ...... *ansatum.*
    Transverse view with three protuberances on each side..... *circulare.*
    Transverse view with four protuberances on each side ...... *Didelta.*
15. End lobe distinct, dilated and joined to the basal portion by a neck. 16
    Margins sinuated but not lobed .... *sublobatum.*
16. Protuberances smooth; end margin straight .... *pectinatum.*
    Protuberances granulated; end margin concave .... *gemmatum.*

8. Cosmarium.

1. Constriction forming a linear notch on each side; end view not circular 2
   Constriction not forming notches at the sides; end view circular.. 24
2. Segments smooth, entire 3
   Segments crenate or margined by pearly granules 9
3. Frond lenticular....................................... *Ralfsii.*
   Frond not lenticular...................................... 4

4. Frond as broad as or broader than long 5
   Frond longer than broad 7
5. Segments reniform; end view with a slight lobe on each side
      at the middle .... *Phaseolus.*
   Segments elliptic in both front and end views................ 6
6. Empty frond reddish; notches closed .... *tinctum.*
   Empty frond hyaline; notches gaping .... *bioculatum.*
7. Ends rounded .... *Cucumis.*
   Ends truncate 8
8. Segments quadrate .... *quadratum.*
   Segments oval with flattened ends .... *pyramidatum.*
   Segments truncato-triangular .... *granatum.*
9. End view elliptic 10
   End view with a lobe at the middle of each side 19
10. Frond smooth, crenate..................................... 11
    Frond rough with pearly granules which give a denticulate
       appearance to the margin 13
11. Sides and end of segments bicrenate .... *Meneghinii.*
    Segments with many crenatures 12
12. Segments with flattened, minutely crenate ends; spines of
       sporangium very short .... *crenatum.*
    Segments with rounded, strongly crenate ends; spines of spo-
       rangium elongated .... *undulatum.*
13. Segments truncato-triangular .... *Botrytis.*
    Segments not in the least triangular 14
14. Frond elliptic; granules large and confined to the margin...... *ovale.*
    Frond not elliptic; granules not confined to the margin 15
15. Frond twice as long as broad, with parallel sides .... *amœnum.*
    Frond not twice as long as broad; sides not parallel........... 16
16. Segments quadrangular .... *conspersum.*
    Segments rounded 17
17. Pearly granules longer than broad .... *Brebissonii.*
    Pearly granules broader than long 18
18. Granules giving a crenate appearance to the margin.. *tetraophthalmum.*
    Granules giving a denticulate appearance to the margin. *margaritiferum.*
19. Segments bordered by papillæ-like granules; end view with
       truncate ends .... *cristatum.*
    Pearly granules neither papillæ-like nor confined to the mar-
       gin; end view with rounded ends 20
20. Segments bordered with denticulated crenatures .... *cœlatum.*
    Segments not crenate 21

21. Segments reniform, three times broader than long .... *commissurale.*
    Segments not reniform, about twice as broad as long 22
22. End margin straight .... *Broomeii.*
    End margin not straight 23
23. Inflation reaching to or beyond the end margin............ *ornatum.*
    Inflation confined to the centre of the segment .... *biretum.*
24. Frond rough with pearly granules which give a denticulate
       appearance to the margin 25
    Frond smooth; margin entire............................. 26
25. Segments spherical .... *orbiculatum.*
    Segments subquadrate .... *cylindricum.*
26. Segments spherical................................. *moniliforme.*
    Segments not spherical 27
27. Segments broader than long (hemispherical or semi-oval) .. *connatum.*
    Segments as long as or longer than broad 28
28. Frond (oval) large, distinctly visible to the naked eye ...... *turgidum.*
    Frond minute, scarcely visible to the naked eye 29
29. Frond fusiform or oblong 30
    Frond not tapering; ends broadly rounded 31
30. Puncta distinct .... *Cucurbita.*
    Puncta wanting, or very indistinct .... *Thwaitesii.*
31. Endochrome in longitudinal fillets .... *curtum.*
    Endochrome not in fillets .... *attenuatum.*

9. Xanthidium.

1. Spines divided at the apex........................... *armatum.*
   Spines subulate 2
2. Segment without central projections, its spines usually four . *? octocorne.*
   Segment with central projections; spines eight or more 3
3. Spines scattered .... *aculeatum.*
   Spines marginal, in pairs 4
4. Segments with a solitary basal spine on each side, the rest
      in pairs .... *cristatum.*
   Segments with basal spines in pairs 5
5. Central projections minute and smooth in end view .... *fasciculatum.*
   Central projections prominent and denticulate in end view .. *Brebissonii.*

10. Arthrodesmus.

Segments elliptic .... *convergens.*
Segments with truncate ends .... *Incus.*

11. STAURASTRUM.

1. Frond neither rough nor spinous; angles in end view broadly rounded.   2
   Frond either rough, spinous, or with elongated angles 5

2. Angles with a minute nipple-like projection 3
   Angles rounded, without a projection 4

3. Sides convex .... *tumidum.*
   Sides concave .... *brevispina* β.

4. Segments elliptic in the front view .... *muticum.*
   Segments semiorbicular in the front view .... *orbiculare.*

5. Frond with awned inflated angles, but without other spines 6
   Frond either awnless or having other spines besides the awns 11

6. Frond rough with minute granules .... *lunatum.*
   Frond smooth 7

7. Segments connected by a ribbon-like band .... *cuspidatum.*
   Segments connected without a band 8

8. Awns, in front view, terminating conical lobes .... *aristiferum.*
   Awns, in front view, not terminating lobes 9

9. Awns straight, rarely converging; segments, if elliptic, not
      turgid .... *dejectum.*
   Awns curved and converging; segments turgid-elliptic 10

10. Segments inflated; awns inconspicuous .... *brevispina.*
    Segments turgid; awns distinct .... *Dickiei.*

11. Frond smooth, glabrous; angles prolonged into processes
       divided at the apex 12
    Frond either rough, spinous or dentate; divided processes,
       if any, rough 13

12. End view with an elongated process from each angle .... *brachiatum.*
    End view with short geminate processes from each lobe .... *læve.*

13. Frond with simple, acute, uniform spines; angles in end
       view broadly rounded and entire......................... 14
    Frond, if spinous, either with some or all of its spines not
       subulate, or with angles not rounded in the end view 16

14. Segments quadrangular in the front view .... *Hystrix.*
    Segments semi-orbicular or reniform in the front view 15

15. Spines conico-triangular; end view of frond with convex sides. *muricatum.*
    Spines hair-like; end view of frond with convex or slightly
       concave sides .... *hirsutum.*
    Spines thorn-like; end view of frond with very concave sides. *teliferum.*

16. Segments quadrangular in the front view............ *quadrangulare.*
    Segments not quadrangular in the front view................. 17

17. Segments in front view with a forked spine on each side, otherwise smooth .... *Avicula.*
    Segments in front view having many spines or none 18

18. Frond smooth; end view acutely triangular, with two acute accessory spines to each angle 19
    Frond rough or spinous; angles in the end view not acute 20

19. Segments in the front view with a forked spine on each side. *monticulosum.*
    Segments in the front view with all their spines slender, simple, and directed outwards .... *pungens.*

20. Frond rough with large granules; end view with five or six rounded and toothed lobes .... *sexcostatum.*
    Frond either minutely granulate or spinous; angles in end view either entire or elongated into rays 21

21. Frond rough with puncta-like granules; angles or rays in end view entire 22
    Frond either spinous or having the angles or rays tipped with spines 28

22. End view with rounded or truncate short rays or angles 23
    End view with elongated slender rays or processes 27

23. End view with slightly concave sides and broadly rounded angles .... *punctulatum.*
    End view either with very concave sides or with angles elongated into short rays 24

24. End view triangular 25
    End view with four or more angles or rays 26

25. Segments rounded on each side .... *alternans.*
    Segments tapering on each side into a short process .... *tricorne.*

26. Segments twice as broad as long; end view 4-rayed ...... *dilatatum.*
    Segments as long as broad; end view 5- to 7-rayed .. *margaritaceum.*

27. Segments with converging processes .... *Arachne.*
    Segments with diverging processes .... *tetracerum.*

28. Frond rough with minute granules; spines, besides those that tip the rays, few and inconspicuous, or none 29
    Frond with evident, more or less forked or emarginate spines 33

29. Processes much elongated 30
    Processes short 31

30. Processes divergent in the front view .... *paradoxum.*
    Processes parallel in the front view .... *gracile.*

31. Processes convergent in the front view .... *cyrtocerum.*
    Processes nearly straight in the front view 32

32. Segments fusiform .... *tricorne β.*
    Segments ventricose, not fusiform .... *polymorphum.*

33. End view with broad, emarginate or bifid lobes .... *enorme.*
    End view with the angles either rounded or tapering into lobes .. 34
34. Spines scarcely distinguishable from the associated granules .. *asperum.*
    Spines distinct 35
35. End view with convex sides and rounded angles, and fringed
    with stout forked spines .... *spongiosum.*
    End view with concave sides; angles not rounded 36
36. Each angle in end view terminated by a spine with two
    accessory spines .... *spinosum.*
    Each angle in end view tapering into a process tipped by
    three or four acute points or spines 37
37. Each side in end view with two slender forked spines, with
    or without other spines; rays slender .... *vestitum.*
    End view without slender forked spines at the sides; rays short, stout. 38
38. Rays curved in end view .... *controversum.*
    Rays straight in end view .... *aculeatum.*

12. Didymocladon.

.... *furcigerus.*

13. Tetmemorus.

1. Frond fusiform in the front view; puncta distinct, scattered. *granulatus.*
   Frond not fusiform in the front view; puncta either very indistinct or in longitudinal rows 2
2. Puncta distinct, in longitudinal rows .... *Brebissonii.*
   Puncta either wanting or indistinct and scattered .... *lævis.*

14. Penium.

1. Frond granulate or striate 2
   Frond smooth 3
2. Granules arranged in longitudinal rows .... *margaritaceum.*
   Granules scattered .... *Cylindrus.*
3. Endochrome transversely divided into four portions .... *interruptum.*
   Endochrome not divided into more than two portions 4
4. Frond lanceolate, fusiform or elliptic 5
   Frond cylindrical 6
5. Frond elliptic; fillets undulating .... *Digitus.*
   Frond lanceolate or fusiform; fillets straight .... *closterioides.*
6. Ends truncate .... *truncatum.*
   Ends rounded 7

7. Sporangium orbicular; conjugated fronds deciduous .... *Jenneri.*
   Sporangium at first quadrate; conjugated fronds persistent. *Brebissonii.*

15. Docidium.

1. Frond neither constricted nor inflated .... ? *asperum.*
   Frond constricted; segments mostly inflated at the base 2

2. Suture inconspicuous; segments not inflated .... *minutum.*
   Suture distinct; segments inflated at their base 3

3. Frond stout; suture projecting at the sides 4
   Frond slender; suture not projecting 5

4. Segments with one inflation; extremities attenuated ...... *truncatum.*
   Segments with several inflations; extremities not attenuated. *nodulosum.*

5. Extremities clavate .... *clavatum.*
   Extremities not clavate 6

6. Ends bordered with tubercles .... *Ehrenbergii.*
   Ends destitute of tubercles .... *Baculum.*

16. Closterium.

1. Frond suddenly constricted at each end into a conical point. *attenuatum.*
   Frond not constricted at the ends 2

2. Frond striated, tapering into a beak at each end, lower
      margin prominent at the middle 3
   Frond not rostrate; if striated, its lower margin not pro-
      minent at the middle 6

3. Beaks setaceous, as long as or longer than the body 4
   Beaks linear, much shorter than the body 5

4. Beaks much longer than the body .... *setaceum.*
   Beaks about as long as the body .... *rostratum.*

5. Frond much inflated at the middle, rapidly tapering at each end. *Ralfsii.*
   Frond slightly inflated at the middle, gradually tapering at
      each end .... *lineatum.*

6. Frond minute, acicular; sporangium cruciform 7
   Frond not acicular; sporangium orbicular 8

7. Ends obtuse .... *Cornu.*
   Ends acute .... *acutum.*

8. Frond semilunate or semilanceolate, the lower margin in-
      clined upwards at the ends 9
   Frond having either truncate ends or its lower margin
      inclined downwards at the ends 12

9. Vesicles numerous, scattered .... *Lunula.*
   Vesicles in a longitudinal row 10

10. Ends of frond slightly curved upwards; striæ distinct .... *turgidum*.
    Ends of frond straight; striæ none or indistinct 11

11. Frond linear-lanceolate; ends conical, obtuse .... *acerosum*.
    Frond semilanceolate; ends subacute .... *lanceolatum*.

12. Frond smooth, crescent-shaped 13
    Frond either not crescent-shaped or else distinctly striated 17

13. Vesicles numerous, scattered .... *Ehrenbergii*.
    Vesicles in a longitudinal row 14

14. Empty frond colourless; ends rounded 15
    Empty frond usually reddish; ends subacute 16

15. Lower margin of frond inflated at the middle .... *moniliferum*.
    Frond not inflated at the middle .... *Jenneri*.

16. Frond inflated at the middle .... *Leibleinii*.
    Frond slender, not inflated at the middle .... *Dianæ*.

17. Lower margin of frond inclined upwards at the truncate
       ends; striæ none or indistinct.................... *didymotocum*.
    Ends of frond inclined downwards; striæ distinct 18

18. Striæ three to seven, prominent 19
    Striæ numerous, fine 20

19. Frond semilunate or crescent-shaped .... *costatum*.
    Frond linear .... *angustatum*.

20. Frond narrow-linear, nearly straight .... *juncidum*.
    Frond tapering, curved 21

21. Striæ crowded; sutures one to three .... *striolatum*.
    Striæ not crowded; sutures usually more than three .... *intermedium*.

17. SPIROTÆNIA.

Spire solitary, broad, conspicuous .... *condensata*.
Spires several, narrow, obscure .... *obscura*.

18. ANKISTRODESMUS.

.... *falcatus*.

19. PEDIASTRUM.

1. Lobes of marginal cells bifid or truncato-emarginate 2
   Lobes of marginal cells entire 3

2. Cells four .... *Tetras*.
   Cells seven, circularly disposed round a central one ...... *Heptactis*.
   Cells eleven, circularly disposed round five or six central ones. *biradiatum*.

3. Marginal cells crescent-shaped; connecting substance coloured. *Selenæa*.
   Cells not crescent-shaped; connecting substance hyaline 4

4. Cells granulate .... *granulatum.*
   Cells smooth 5
5. Lobes of marginal cells terminated by obtuse processes .... *ellipticum.*
   Lobes of marginal cells acute 6
6. Marginal cells with notches as wide as the cells .... *pertusum.*
   Notches much narrower than the cells 7
7. Marginal cells four .... *simplex.*
   Marginal cells more than four 8
8. Six cells circularly disposed round two central ones ...... *Napoleonis.*
   Cells arranged in two or more circles 9
9. Lobes of marginal cells angular, not tapering .... *angulosum.*
   Lobes of marginal cells tapering .... *Boryanum.*

20. SCENEDESMUS.

1. Cells obtuse 2
   Cells acute or acuminate 3
2. Cells cylindrical or oblong .... *quadricauda.*
   Cells ovate .... *obtusus.*
3. Cells ventricose at the middle 4
   Cells slender, not at all ventricose 5
4. Cells in a single zigzag row .... *acutus.*
   Cells either in two rows or not zigzag .... *obliquus.*
5. Outer cells crescent-shaped, inner straight, all approximate. *dimorphus.*
   Cells curved, distant .... *duplex.*

APPENDIX.

LIST OF SPECIES NOT HITHERTO DETECTED IN BRITAIN.

HYALOTHECA, *Ehr*.

3. *H.? dubia* (Kütz.); filament without a mucous tube; joints rather broader than long, with two puncta near each margin.

Hyalotheca? dubia, Kützing, *Phycologia Germanica*, p. 140 (1845); *in lit. cum icone.*

Nordhausen, Prussia, *Kützing.*

Tab. XXXV. fig. 16. portion of a filament from a drawing by Professor Kützing.

DESMIDIUM, *Ag*.

3. *D. undulatum* (Corda); joints with four crenatures at each margin.

Desmidium undulatum, Corda, *Observations Microscopiques sur les Animalcules des eaux et des thermes de Carlsbad*, p. 19. t. 4. f. 27 (1840).

Carlsbad; Prague; and Reichenberg, *Corda.*

The constriction of the joints is marked by slight marginal notches, on each side of which are two broad crenatures.

Desmidium undulatum appears to be a distinct species, differing from *D. Swartzii* by having bicrenate segments.

4. *D. didymum* (Corda).

Desmidium didymum, Corda, *Alm. de Carlsbad* 1835, p. 123. t. 4. f. 43.
Kützing, *Phyc. Germ.* p. 141.

Carlsbad, *Corda.*

Ehrenberg and Meneghini unite *Desmidium didymum* (Corda) to *D. bifidum* (Ehr.), which the latter describes as a filamentous plant. Kützing removes *D. bifidum* to *Staurastrum*, but retains the present plant in *Desmidium*, of which Corda considers that both are true species. I have not seen Corda's descriptions or figures, but in a subsequent work he gives the following reasons under *D. bifidum* for believing them distinct:—

"M. Ehrenberg lui adjoint l'espèce suivante, trouvée à Carlsbad. Elle en diffère totalement, du moins de celle que je considère comme la forme normale du *Desmidium bifidum*, qui, relativement à sa structure, principalement à la découpure de ses pointes, correspond beaucoup avec le dessin incomplet d'un membre séparé, qu'on trouve dans l'ouvrage de M. Ehrenberg, *l. c. D. bifidum* (*bidens* dans le dessin), vu latéralement, ne présente que deux pointes principales sur chaque aîle, lesquelles sont simples. Quant à l'espèce suivante, nous avons dépeint deux pointes principales, très-visibles, et chacune d'elles également pourvue de deux pointes."—*Observ. Microscop. sur les Animalcules de Carlsbad*, p. 18.

5. *D. bifidum* (Ehr.). See *Staurastrum bifidum*.

APTOGONUM.

2. *A. Baileyi* ——; filament not crenated; joints about equal in length and breadth.

Odontella? tridentata, Bailey, *in lit. cum icone* (1846).

Worden's Pond, Rhode Island; near Princeton, New Jersey, with sporangia, *Bailey*.

Filament triangular; joints excavated at their junction like those of *Aptogonum Desmidium*. The joints are not bicrenate, hence the margins of the filament are entire, a character which distinguishes it from that species. The end view is triangular, with rounded angles.

Professor Bailey has sent me a drawing of the conjugated state, very interesting from its resemblance to a condition of *Desmidium Swartzii*, which I had doubtfully regarded as the sporangia of that species. In both plants it is difficult to understand the process or to distinguish the coupled filaments, since the appearance is merely that of a much-enlarged and torn filament.

Tab. XXXV. fig. 1. from drawings by Professor Bailey; *a*. portion of a filament; *b*. transverse view; *c*. sporangia highly magnified

SPHÆROZOSMA, Corda.

3. *S. lamelliferum* (Corda).

Sphærozosma lamelliferum, Corda, *Observ. Microscop. sur les Animalc. de Carlsbad*, p. 21. t. 4. f. 29 (1840).

Carlsbad, *Corda*.

"Chaînes courtes, les membres doubles presque uniformes; un tiers est découpé transversalement des deux côtés; lobes rapprochés l'un de l'autre, arrondis; cuirasse unie, transparente et blanche. Le membre de réunion feuilleté, blanc."—*Corda, l. c.*

4. *S. pulchrum* (Bailey); joints twice as broad as long, deeply incised on each side; junction-margins straight, connected by short bands.

Sphærozosma pulchrum, Bailey, *in lit. cum icone* (1847).

West Point, New York; and Princeton, New Jersey, *Bailey.*

Professor Bailey informs me that this species is twice as large as *Sphærozosma vertebratum*. His figure represents the filament diminishing and increasing in breadth at intervals, an appearance probably caused by the twisting of the plant. *Sphærozosma pulchrum* agrees with *S. vertebratum* in the presence of a mucous tube, and also in the form of the joints; but not in the mode of their connexion with each other. It differs from Corda's figure of *S. lamelliferum*, inasmuch as its segments are not reniform.

Tab. XXXV. fig. 2. *a.* portion of a filament; *b.* joints with mucous tube, from drawings by Professor Bailey.

5. *S.? filiforme* (Ehr.); joints bilobed, united by double slender processes which inclose a quadrate foramen between each pair.

Odontella? filiformis, Ehrenberg, *Infusor.* p. 154 (1838); *Meteorp.* t. 1. f. 20.
Tessarthra filiformis, Ehrenberg, *Infusor.* t. 10. f. 21 (1838).
Isthmia filiformis, Meneghini, *Synop. Desmid. in Linnæa* 1840, p. 205.
Isthmosira filiformis, Kützing, *Phycologia Germanica*, p. 141 (1845).

Germany, *Ehrenberg, Kützing.*

MICRASTERIAS, *Ag.*

14. *M. apiculata* (Ehr.); frond orbicular, rough with scattered spines; segments five-lobed, lobes incised and toothed, end lobe narrow.

Euastrum apiculatum, Ehrenberg, *Abh. der Berl. Ak.* 1833, p. 245; *Infusor.* p. 167. Kützing, *Phyc. Germ.* p. 134.
Euastrum aculeatum, Ehrenberg, *Infus.* t. 12. f. 2 (1838).
Micrasterias apiculata, Meneghini, *Synop. Desmid. in Linnæa* 1840, p. 216. Brébisson, *in lit. cum specimine.*

Germany, *Ehrenberg.* Falaise, *Brébisson.*

Specimens sent me by M. de Brébisson have deeper incisions than that figured by Ehrenberg, and are also smaller.

15. *M. quadragies-cuspidata* (Corda); frond with scattered hair-like spines; end lobe the broadest.

Cosmarium quadragies-cuspidatum, Corda, *Observ. Microscop. sur les Animalc. de Carlsbad*, p. 30. t. 6. f. 40 (1840).

Carlsbad, *Corda.*

This plant bears the same relation to *Micrasterias apiculata* which *M. crenata* and *M. truncata* bear to *M. denticulata* and *M. rotata*. I subjoin the following description from Corda's work:—

"Corpuscules chevelus, verts, presque à six lobes. Lobe du milieu triangulaire, les coins extérieurs à deux pointes; au milieu, deux cylindres vert foncé, appuyés l'un contre l'autre. Chaque lobe latéral a trois découpures, et chaque découpure a deux dents."

16. *M. foliacea* (Bailey); frond subquadrate; end lobes narrow, with emarginate angles; lateral lobes inciso-dentate, with a short rounded tooth-like projection next the end lobe.

Micrasterias foliacea, Bailey, *in lit. cum icone* (1847).

Worden's Pond, Rhode Island, *Bailey*.

"Frond small, smooth, nearly quadrangular in the general outline: segments composed of three principal lobes, of which the middle one is nearly simple and emarginate: the lateral lobes are incised, in much the same manner as in *Micrasterias rotata*, and all their subdivisions extend to an equal distance from the middle line of the frond."—*Bailey, in lit.*

Tab. XXXV. fig. 3. Frond dividing, from a drawing by Professor Bailey.

17. *M. Torreyi* (Bailey); frond orbicular; lateral lobes deeply incised; inner subdivisions acute, external bidentate at the apex, all tapering.

Micrasterias Torreyi, Bailey, *in lit. cum icone* (1847).

Near Princeton, New Jersey, *Bailey*.

This species connects the elliptic with the orbicular species. The end lobe is narrow and not exserted; the angles taper into acute points or spines. All the incisions are deep. The lateral lobes have five subdivisions, two belonging to the basal lobe and three to the intermediate; but the distinction is obscure, and only indicated by the rather deeper incision which parts them. Of the subdivisions, that next the other segment and that adjacent to the end lobe are bidentate at the apex, the rest taper gradually into acute points.

Tab. XXXV. fig. 5. Frond from a drawing by Professor Bailey.

18. *M. muricata* (Bailey); "segments divided by deep indentations into three transverse portions; the basal with three linear processes on each side, the others with two on each side."

Euastrum muricatum, Bailey, *American Journal of Science and Arts*, 1846, p. 126. figs. 1, 2.

Catskill Mountains, United States, *Bailey*.

Micrasterias muricata belongs to the same section as *M. furcata*, but differs from that and every other known species in some remarkable particulars. The division into five lobes is indicated merely by the presence of processes, which, unlike those in the other species, do not diverge in the front view, but spread laterally, in such a manner that the one nearest the eye more or less conceals its companions.

19. *M. Baileyi* ——; frond granulated; segments three-lobed; lobes bipartite, end one much-exserted.

Micrasterias, n. sp., Bailey, *in lit. cum icone* (1846).

West Point, New York; and Rhode Island, *Bailey*.

Micrasterias Baileyi is allied to *M. furcata* and *M. Crux-Melitensis*; the frond however is minutely granulate, the segments are only three-lobed, and the end lobe is more exserted than it is in those species. The basal lobes are deeply bipartite, the end one merely sinuated. All the subdivisions are bidentate at the apex.

Tab. XXXV. fig. 4. Empty frond: from a drawing by Professor Bailey.

20. *M. incisa* (Kütz.); lobes horizontal, basal ones truncate, with a tooth at each angle; end lobe convex, its angles acute.

Euastrum Crux-Melitensis, Ehrenberg, *Infus.* t. 12. f. 3 c. (1838).
Euastrum incisum, Brébisson, Meneghini, *Synop. Desmid. in Linnæa* 1840, p. 216. Kützing, *Phyc. Germ.* p. 134.
Micrasterias incisa, Brébisson, *in lit.* (1846).

Germany, *Ehrenberg*. Falaise, *Brébisson*.

EUASTRUM, *Ehr.*

19. *E. cornutum* (Kütz.); segments three-lobed; terminal lobe cuneate, included between two process-like projections of the basal portion.

Euastrum cornutum, Kützing, *Phycologia Germanica*, p. 135 (1845); *in lit. cum icone.*

Germany, *Kützing*.

The frond resembles that of *Euastrum crassum*, but the outer angles of the basal portion of each segment are prolonged into rostrate-like processes which include the terminal lobe between them.

Tab. XXXV. fig. 13. Frond: from a drawing by Professor Kützing.

20. *E. Pelta* (Corda); segments quadrangular; end margin with a rounded protuberance at each corner; lateral margins with a small protuberance at the basal end and a larger one near the outer end.

Cosmarium Pelta, Corda, *Almanach de Carlsbad* 1835, p. 121. t. 2. f. 25. Meneghini, *Synop. Desmid. in Linnæa* 1840, p. 222.
Euastrum Pelta, Kützing, *Phycologia Germanica*, p. 135 (1845).

Carlsbad, *Corda*.

21. *E. crenatum* (Kützing).

Euastrum crenatum, Kützing, *Phycologia Germanica*, p. 135 (1845).

Dalmatia, *Kützing*.

This plant may be a variety of *Euastrum elegans*.

Tab. XXXV. fig. 14. Frond: from a drawing by Professor Kützing.

22. *E. crenulatum* (Ehr.).

Euastrum crenulatum, Ehrenberg, *Meteorp.* t. 1. f. 16. Kützing, *Phycologia Germanica*, p. 135.

Germany, *Ehrenberg.*

I am unacquainted with the characters of this species.

COSMARIUM, *Corda.*

34. *C. Palangula* (Bréb.).

Cosmarium Palangula, Brébisson, *in lit. cum icone* (1846).

Falaise, *Brébisson.*

M. de Brébisson informs me that " Le *Cosmarium Palangula* me semble se distinguer du *Cos. Cucurbita* par sa forme plus cylindrique, tronquée et non en *Courge*, ses séries de points plus fines, transverses et non longitudinales." I am however unable to distinguish this species from *C. Cucurbita*, for the puncta are scattered in British specimens of the latter.

35. *C. lagenarium* (Corda); segments triangular, all the angles broadly rounded.

Cosmarium lagenarium, Corda, *Alm. de Carlsbad* 1835, p. 121. t. 2. f. 26. Brébisson, *in lit.*

Carlsbad, *Corda.*

36. *C. Papilio* (Menegh.); segments smooth, triangular, with rectangular apex; end view linear, with a lobe at the middle of each side.

Cosmarium Papilio, Meneghini, *Consp. Alg. Eugan.* p. 18 (1837); *Synop. Desmid. in Linnæa* 1840, p. 223.

Italy, *Meneghini.*

37. *C. ventricosum* (Kütz.).

C. ventricosum, Kützing, *Phycologia Germanica*, p. 136 (1845); *in lit. cum icone.*

I believe this to be a species of *Cosmarium* in a dividing state.

XANTHIDIUM, *Ehr.*

7. *X. Artiscon* (Ehr.); segments narrowed at the base; end margin with numerous elongated spines which are divided at the apex.

Xanthidium, No. 2, Bailey, *American Journal of Science and Arts*, v. 41. p. 291. t. 1. f. 15 (1841).
Xanthidium Artiscon, Ehrenberg, *Verbreitung und Einfluss des Mikroskopischen Lebens in Süd- und Nord-Amerika* (1843).

West Point, New York, *Bailey.*

Xanthidium Artiscon differs from *X. armatum* by its segments tapering at the base: its spines also are much longer and are more restricted to the outer and rounded margin.

Several plants referred by Ehrenberg to *Xanthidium* are angular in the end view, and properly belong to *Staurastrum*.

8. *X. furcatum* (Ehr.); "corpuscules globose, green, single or binate, spinous; spines scattered, forked at the apex."

Xanthidium furcatum, Ehrenberg, *Infus.* p. 148. t. 10. f. 25 (1838); *Meteorp.* t. 1. f. 21.

This plant is probably a *Staurastrum*. M. de Brébisson refers it to *S. spinosum*, and Professor Kützing makes it a synonym of his *Phycastrum furcigerum* (*Didymocladon furcigerus*). Of neither have I seen specimens resembling Ehrenberg's figure, from which indeed I believe the latter plant is very distinct.

ARTHRODESMUS, *Ehr.*

3. *A. minutus* (Kütz.).

Euastrum Incus, Kützing, *Phycologia Germanica*, p. 137 (1845); (not *Staurastrum Incus*, Meneghini).
Euastrum minutum, Kützing, *in lit. cum icone* (1846).
Arthrodesmus minutus, Brébisson, *in lit.* (1846).

Germany, *Kützing*.

Tab. XXXV. fig. 15. *a, b*. perfect fronds; *c*. dividing frond: from drawings by Professor Kützing.

4. *A.? truncatus* (Ehr.); "corpuscules green, slightly compressed, campanulate, geminate, externally truncate, spinous."

Arthrodesmus truncatus, Ehrenberg, *Infus.* p. 152 (1838).

This plant is probably either a *Xanthidium* or a *Staurastrum*.

STAURASTRUM, *Meyen.*

39. *S. pygmæum* (Bréb.); segments cuneiform; end view triangular, with slightly rounded sides.

Staurastrum pygmæum, Brébisson, *in lit. cum icone* (1846).

Falaise, *Brébisson*.

Staurastrum pygmæum is smaller than *S. alternans*, its sides, in the end view, are more convex, and its angles less rounded. The sporangium is orbicular and spinous.

Tab. XXXV. fig. 26. from drawings by M. de Brébisson: *a*. front view of frond; *b*. immature sporangium; *c*. mature sporangium.

40. *S. rugulosum* (Bréb.); segments elliptic, denticulate at their sides; end view triangular, with the angles broadly rounded and denticulate.

Euastrum, No. 9, Bailey, *American Journ. of Science and Arts*, v. 41. p. 296. t. 1. f. 9 (1841).
Staurastrum rugulosum, Brébisson, *in lit. cum icone* (1846).

West Point, New York, *Bailey*. Falaise, *Brébisson*.

Tab. XXXV. fig. 19. from drawings by M. de Brébisson: *a.* front view; *b.* end view.

41. *S. scabrum* (Bréb.); segments elliptic, scabrous; end view triangular, fringed with minute emarginate spines.

Staurastrum scabrum, Brébisson, *in lit. cum icone* (1846).

Falaise, *Brébisson*.

Tab. XXXV. fig. 20. from drawings by M. de Brébisson: *a.* front view; *b.* end view.

42. *S. bacillare* (Bréb.); frond smooth; processes capitate; end view with three to five capitate rays or processes.

Binatella bacillaris, Brébisson, *Alg. Fal.* p. 66 (1835).
Staurastrum bacillare, Bréb., Meneghini, *Synop. Desmid. in Linnæa* 1840, p. 228. Brébisson, *in lit. cum icone*.

Falaise, *Brébisson*.

Frond smooth; processes in front view stout, divergent, each terminated by a round knob; end view three- to five-rayed, the rays capitate in the same manner as in the front view.

A remarkable and very distinct species.

Tab. XXXV. fig. 21. from drawings by M. de Brébisson: *a.* front view of frond; *b, c, d.* transverse views.

43. *S. Capitulum* (Bréb.); segments quadrate, sinuated on each side, prominences rough; end view triangular, with broadly rounded angles.

Staurastrum Capitulum, Brébisson, *in lit. cum icone* (1836).

Falaise, *Brébisson*.

Frond quadrilateral, nearly twice as long as broad; constriction shallow; the segments have two rounded prominences on each side and a small semicircular sinus between them; these prominences are rough with minute granules, which give them a crenate appearance. End margin straight. End view triangular, the angles broadly rounded; sides nearly straight, but each slightly constricted at the middle.

Tab. XXXV. fig. 25. from drawings by M. de Brébisson: *a.* front view of frond; *b.* end view.

44. *S. pileolatum* (Bréb.); frond quadrilateral, slightly constricted at the middle; segments terminated by three conical processes.

Staurastrum pileolatum, Brébisson, *in lit. cum icone* (1846).

Falaise, *Brébisson*.

Frond twice as long as broad; segments quadrate, with a broad, shallow sinus on each side; the end terminated by three conspicuous conical processes, which are rough with minute granules. End view triangular, the angles rounded.

Tab. XXXV. fig. 22. from drawings by M. de Brébisson: *a.* front view of frond; *b.* end view.

45. *S. echinatum* (Bréb.).

Staurastrum echinatum, Brébisson, *in lit. cum icone* (1846).

Falaise, *Brébisson*.

This species appears, from M. de Brébisson's drawings, to be closely allied to *Staurastrum hirsutum* and *S. teliferum*.

Tab. XXXV. fig. 24. from drawings by M. de Brébisson: *a.* front view of frond; *b.* end view.

46. *S. crenatum* ——; segments cuneate; outer margins crenate; end view with three truncate and crenate angles.

Staurastrum, new sp., Bailey, *in lit. cum icone* (1847).

Rhode Island, *Bailey*.

Tab. XXXV. fig. 17. *a.* front view of frond; *b.* end view: from drawings by Professor Bailey.

47. *S. bifidum* (Ehr.); frond smooth; end view with three cloven angles.

Desmidium bifidum, Ehr. *Abh. der Berlin Ak.* 1832, p. 292; *Infusor.* p. 141. t. 10. f. 11.
Phycastrum bifidum, Kützing, *Phycologia Germanica*, p. 138 (1845).
Staurastrum bifidum, Brébisson, *in lit.* (1846).

Germany, *Ehrenberg*.

48. *S. eustephanum* (Ehr.); end view triangular with six emarginate spines on the upper surface; each angle terminated by a short ray tipped with spines.

Desmidium eustephanum, Ehrenberg, *Verbr. und Einfluss des Mikrosk. Lebens in Süd- und Nord-Amerika*, t. 4. f. 23 (1843).

West Point, New York, *Bailey*.

49. *S. senarium*, Ehr.

Desmidium senarium, Ehrenberg, *Verbr. und Einfluss des Mikrosk. Lebens in Süd- und Nord-Amerika*, t. 4. f. 22 (1843).

West Point, New York, *Bailey*.

Ehrenberg's figure represents the end view as triangular, the angles terminating in short rays, which are tipped by minute spines: on each side are two short forked spines, and six others on the upper surface. His figure agrees in some respects with *Staurastrum spinosum*, consequently *S. senarium* may prove a variety of that very variable species.

50. *S. Ehrenbergii*, Corda.

Xanthidium Ehrenbergii, Corda, *Observ. Microscopiques sur les Animalcules de Carlsbad*, p. 29. t. 5. f. 36, 37 (1840).

Prague, *Corda*.

"Corpuscules par paire, vus de côté, ovales; vus d'en haut, triangulaires, munis de six appendices terminaux et latéraux, et de deux autres appendices centraux, qui sont courts, blancs, en fourchette, mais à pointes divergentes."— Corda, *l. c.*

Corda's figure of the front view resembles *Staurastrum spinosum*.

51. *S. articulatum*, Corda.

Xanthidium articulatum, Corda, *Observ. Microscopiques sur les Animalcules de Carlsbad*, p. 28. t. 5. f. 35 (1840).

Carlsbad, Prague and Reichenberg, *Corda*.

"Corpuscules ovales, par paire, munis aux deux bouts d'un appendice à deux cellules, qui se divise encore en forme de fourchette; et latéralement en deux appendices plus longs à quatre cellules, et une pointe en fourchette. Sur les deux côtés plats, se trouvent deux protubérances transversales, également pourvues de deux allongements cellulaires en fourchette."—Corda, *l. c.*

52. *S. coronatum* (Ehr.); end view triangular and terminating at each angle in three short, diverging arms which are divided at the apex.

Xanthidium, No. 3, Bailey, *American Journ. of Science and Arts*, v. 41. p. 291. t. 1. f. 16 (1841).
Xanthidium coronatum, Ehrenberg, *Verbr. und Einfluss des Mikrosk. Lebens in Süd- und Nord-Amerika*, t. 4. f. 26 (1843).

West Point, New York, *Bailey*.

Professor Bailey's figure, which in a list of American Desmidieæ obligingly communicated he refers to as *Xanthidium coronatum*, Ehr., in the front view resembles *Xanthidium Artiscon*, whilst Ehrenberg's appears to me a good representation of the front view of *Didymocladon furcigerus*. I doubt therefore whether these two figures can represent the same species.

53. S.? *minus* (Kütz.); frond smooth; end view with five slender acute rays.

Pentasterias minor, Kützing, *Phycologia Generalis*, p. 162 (1843).
Phycastrum? minus, Kützing, *Phycologia Germanica*, p. 139 (1845); *in lit. cum icone.*

Nordhausen, Prussia, *Kützing.*

Tab. XXXV. fig. 18. *a, b.* Fronds: from drawings by Professor Kützing.

54. S. *glabrum* (Ehr.); frond smooth; end view triangular, each angle terminated by a mucro-like spine.

Euastrum, No. 14, Bailey, *Amer. Journ. of Science and Arts*, v. 41. p. 297. t. 1. f. 14 (1841).
Desmidium glabrum, Ehrenberg, *Verbreitung und Einfluss des Mikrosk. Lebens in Süd- und Nord-Amerika* (1843); *Meteorp.* t. 1. f. 3.
Phycastrum glabrum, Kützing, *Phycologia Germanica*, p. 137 (1845).

West Point, New York, *Bailey.* Germany, *Ehrenberg.*

This species is probably identical with *Staurastrum Avicula*.

55. S. *granulosum* (Ehr.).

Desmidium granulosum, Ehrenberg, *Meteorp.* t. 1. f. 12.
Phycastrum granulosum, Kützing, *Phycologia Germanica*, p. 138 (1845).

Germany, *Ehrenberg.*

This species is known to me only by name.

56. S. *globulatum* (Bréb.); segments fusiform, capitate; end view lar, each angle terminated by a granulated knob.

Staurastrum globulatum, Brébisson, *in lit. cum icone* (1846).

Falaise, *Brébisson.*

Tab. XXXV. fig. 23. *a.* end view of frond; *b.* front view: from drawings by M. de Brébisson.

DOCIDIUM, Bréb.

8. D. *coronatum* (Bréb.); frond stout; suture projecting on each side; segments inflated at the base and bordered by tubercles at the end.

Docidium coronatum, Brébisson, *in lit. cum icone* (1846).

Falaise, *Brébisson.*

This species agrees with *Docidium nodulosum* not only in size and form, but in having a suture projecting on each side, and also in its segments, which have undulated margins; but it is distinguished from that plant by having the ends bordered with minute but distinct tubercles. The end view is circular, the margin crenate. *Docidium coronatum* differs from *D. Ehrenbergii* in its larger size and projecting suture.

Tab. XXXV. fig. 6. *a.* empty segment; *b.* end view: from drawings by M. de Brébisson.

9. *D. nodosum* (Bailey); frond stout; segments with four prominent nodes separated by constrictions; end view crenate.

Closterium nodosum, Bailey, *Amer. Journ. of Science and Arts*, 1846, v. 1. p. 127. f. 3.
Docidium nodosum, Bailey, *in lit. cum icone* (1847).

United States, *Bailey*.

An end view shows that each node is not a simple swelling, but really formed by whorls of tubercles.

"This species is easily recognised by the deep indentations in its outline, corresponding to the constrictions which separate the transverse rows of knot-like projections. It is one of the largest species in the genus."—Bailey, *l. c.*

Tab. XXXV. fig. 8. *a.* empty segment; *b.* end view: from drawings by Professor Bailey.

10. *D. constrictum* (Bailey); frond stout; segments with moderately deep constrictions, which separate four equal, gently curving prominences; end view entire.

Docidium constrictum, Bailey, *in lit. cum icone* (1847).

Worden's Pond, Rhode Island, *Bailey*.

"This species is at once distinguished from *Docidium nodosum* by the cross section of the nodes being a simple circle instead of an indented one."—Bailey, *in lit.*

Tab. XXXV. fig. 7. from drawings by Professor Bailey: *a.* front view; *b.* end view.

11. *D. verrucosum* (Bailey); segments with numerous whorls of small prominences which give the margins an undulated appearance, all the undulations equal.

Closterium verrucosum, Bailey, *American Journal of Science and Arts*, 1846, v. 1. p. 127. f. 4.

New York; Rhode Island, *Bailey*.

"This is a very pretty species with a waved outline, caused by the slight projections, which are arranged in numerous transverse rings."—Bailey, *l. c.*

12. *D. verticillatum* (Bailey); segments with numerous whorls of tooth-like projections; ends with three bidentate processes.

Docidium verticillatum, Bailey, *in lit. cum icone* (1847).
Triploceras verticillatum, Bailey, *in lit.* (1847).

Worden's Pond, Rhode Island; Princeton, New Jersey, *Bailey*.

A most remarkable and interesting plant, for a specimen of which I am indebted to Professor Bailey.

The triple terminal processes are so unlike what we find in other species of *Docidium*, that Professor Bailey, in a letter to me, proposes to form a new genus for the reception of this plant.

Tab. XXXV. fig. 9. *a.* empty frond; *b, c.* empty segments; *d, e.* extremities highly magnified: from drawings by Professor Bailey.

13. *D. crenulatum* (Ehr.).

Closterium crenulatum, Ehrenberg, *Verbreitung und Einfluss des Mikrosk. Lebens in Süd- und Nord-Amerika* (1843).

Professor Bailey informs me that this plant is identical with *Docidium nodulosum*.

CLOSTERIUM, *Nitzsch*.

23. *C. cuspidatum* (Bailey); frond smooth, crescent-shaped; ends mucronate.

Closterium cuspidatum, Bailey, *in lit. cum icone* (1847).

Worden's Pond, Rhode Island, *Bailey*.

Closterium cuspidatum differs from every other species of the genus in having each end tipped by a spine or mucro.

Tab. XXXV. fig. 11. Frond: from a drawing by Professor Bailey.

24. *C. Cucumis* (Ehr.); frond smooth, stout, semilunate; ends broadly rounded.

Closterium Cucumis, Ehrenberg, *Verbreitung und Einfluss des Mikrosk. Lebens in Süd- und Nord-Amerika*, t. 4. f. 29 (1843).

New York, *Bailey*.

Ehrenberg's figure represents this species about five times longer than broad, in form resembling *Closterium Lunula*, but stouter in proportion to its length, and having more rounded ends.

25. *C. Thureti* (Bréb.); frond smooth, crescent-shaped; ends subacute; margins unconnected at the suture; vesicles in a single series.

Closterium Thureti, Brébisson, *in lit. cum icone* (1846).

Falaise, *Brébisson*.

26. *C. arcuatum* (Bréb.); frond smooth, slender-crescent-shaped; ends obtuse, scarcely notched.

Closterium arcuatum, Brébisson, *in lit. cum icone* (1845).

Closterium acuminatum, Kützing, *Phycologia Germanica,* p. 130 (1845), (according to Kützing *in lit.*).

Falaise, *Brébisson.* Germany, *Kützing.*

I am unable to distinguish *Closterium arcuatum* from *C. Dianæ*; but in drawings sent me by M. de Brébisson, the latter has distinct terminal notches, which are wanting in the former.

27. *C. Venus* (Kütz.).

Closterium Venus, Kützing, *Phycologia Germanica,* p. 130 (1845); *in lit. cum icone.*

Germany, *Kützing.*

I am indebted to Professor Kützing's kindness for a drawing of this species, but I must confess that I am unable to discover any characters sufficient to distinguish it from *Closterium Dianæ*.

Tab. XXXV. fig. 12. Frond, from a drawing by Professor Kützing.

28. *C. Amblyonema* (Ehr.); frond stout, linear, slightly curved, twenty times longer than broad; ends rounded.

Closterium lineatum, Bailey, *American Journ. of Science and Arts,* v. 41. p. 303. t. 1. f. 34 (1841). ?

Closterium Amblyonema, Ehrenberg, *Verbreitung und Einfluss des Mikrosk. Lebens in Süd- und Nord-Amerika* (1843).

West Point, New York, *Bailey.*

29. *C. uncinatum* (Kütz.); frond slender, finely and closely striated; extremities tapering to a subacute point and suddenly curved downwards.

Closterium uncinatum, Kützing, *Phycologia Germanica,* p. 131 (1845); *in lit. cum icone.*

Germany, *Kützing.*

30. *C. decussatum* (Kütz.); frond stout, finely and closely striated, slightly curved, gradually tapering; extremities slender, but obtuse at the apex.

Closterium decussatum, Kützing, *Phycologia Germanica,* p. 131 (1845); *in lit. cum icone.*

Germany, *Kützing.*

A drawing sent me by Professor Kützing represents the striæ as regularly crossing each other, so as to form diamond-shaped reticulations. As this appearance is not unusual in dried specimens, when the flattened frond permits the striæ of both surfaces to be visible together, I will venture to suggest the possibility that Professor Kützing's drawing may have been taken from a frond in that condition. *Closterium decussatum* seems to differ from *C. turgidum* in its more tapering extremities.

31. *C. turgidulum* (Kütz.); frond stout, curved; extremities slender, gradually tapering; striæ few, conspicuous.

Closterium turgidulum, Kützing, *Phycologia Germanica*, p. 132 (1845); *in lit. cum icone.*

Germany, *Kützing*.

Closterium turgidulum differs from *C. costatum* in its more elongated extremities.

32. *C. obtusangulum* (Corda).

Closterium obtusangulum, Corda, *Observations Microscop. sur les Animalc. de Carlsbad*, p. 35. t. 6. f. 42 (1840).

Carlsbad, *Corda*.

" Corps courbé en croissant, quadrangulaire; coins ronds. Cuirasse unie, blanche, arrondie vers les pointes, et transparente; au milieu un ruban transversal étroit, blanc et transparent. Pas de vésicules rotatoires dans les pointes, mais beaucoup de corpuscules rotatoires, ovales, transparents, blancs, se mouvant vivement."—Corda, *l. c.*

33. *C. inæquale* (Ehr.); minute, semilunate; extremities unequal, conic, ends acute; striæ prominent; vesicles scattered.

Closterium inæquale, Ehrenberg, *Abh. der Berlin Ak.* 1831, p. 67; *Infus.* p. 98. t. 6. f. 11. Meneghini, *Synopsis Desmid. in Linnæa* 1840, p. 235. Kützing, *Phycologia Germanica*, p. 131.

Germany, *Ehrenberg, Kützing*.

34. *C. quadrangulare* (Corda).

Closterium quadrangulare, Corda, *Observations Microscop. sur les Animalc. de Carlsbad*, p. 35. t. 6. f. 46 (1840).

Carlsbad; Reichenberg, *Corda*.

" Long, corps mince, quadrangulaire. Coins aigus. Cuirasse unie, transparente, blanche. Substance interne vert foncé, à petits grains. Sur les pointes les quatre surfaces de la cuirasse sont obliquement coupées, et la pointe arrondie. Elle contient de grosses vésicules rotatoires, bien limitées, d'un rouge pâle, avec plusieurs corpuscules rotatoires elliptiques et rougeâtres."—Corda, *l. c.*

35. *C. gracile* (Bréb.); frond slender, smooth, lanceolate, gradually tapering into short beaks, which are curved downwards.

Closterium gracile, Brébisson, Meneghini, *Synop. Desmid. in Linnæa* 1840, p. 234, under *C. lineatum*.
Stauroceras gracile, Brébisson, *in lit. cum icone* (1846).

Falaise, *Brébisson*.

Closterium gracile differs from the other rostrate species in its smooth frond. Its beaks are shorter than those of *C. rostratum* and *C. setaceum*, and its frond is less inflated.

36. *C. tenerrimum* (Kütz.).

Closterium tenerrimum, Kützing, *Phycologia Germanica*, p. 130 (1845); *in lit. cum icone.*

Nordhausen, *Kützing.*

I can perceive in Professor Kützing's drawing no character by which this plant can be separated from *Closterium acutum*.

Tab. XXXV. fig. 10. Fronds: from a drawing by Professor Kützing.

ANKISTRODESMUS, *Corda.*

2. *A. fusiformis* (Corda).

Ankistrodesmus fusiformis, Corda, *Almanach de Carlsbad* 1838, p. 199. t. 2. f. 18.

Carlsbad, *Corda.*

3. *A. convolutus* (Corda).

Ankistrodesmus convolutus, Corda, *Almanach de Carlsbad* 1838, p. 199. t. 2. f. 19.

Carlsbad, *Corda.*

I am unacquainted with the characters of these species.

SCENEDESMUS, *Meyen.*

7. *S. antennatus* (Brèb.); cells fusiform, somewhat ventricose at the middle; ends cuspidate, each terminated by a minute orbicular globule.

Scenedesmus antennatus, Brébisson, *in lit. cum icone* (1846).

Falaise, *Brébisson.*

Scenedesmus antennatus resembles *S. obliquus* in form, and also in the arrangement of its cells; but it is distinguished from that and every other species by having the attenuated points tipped by minute globules.

Tab. XXXV. fig. 27. *a, b.* different states of frond: from drawings by M. de Brébisson.

INDEX.

The names printed in inalics are *synonyms*.

	Page		Page		Page
Achnanthes.		attenuatum	169	rostratum, Ralfs	174
dimorpha, Turpin	191	*Baculum*, Bréb.	158	*rostratum*, Bailey	176
obliqua, Turpin	192	*Baillyanum*, Bréb.	169	*ruficeps*, Ehr.	168
octalterna, Turpin	193	*Brebissonii*, Menegh.	145	*Sceptrum*, Kütz.	158
quadralterna, Turp.	193	*caudatum*, Corda	175	setaceum	176
quadricauda, Turp.	190	*clandestinum*, Kütz.	109	striolatum	170
quadrijuga, Turpin	190	Cornu	176	*subrectum*, Bréb.	169
Ankistrodesmus, 179,	205	costatum	170	*sulcatum*, Bréb.	172
convolutus	222	*crenulatum*, Ehr.	219	tenerrimum	222
falcatus	180	Cucumis	219	*tenue*, Kütz.	176
fusiformis	222	*curtum*, Bréb.	109	*tenue*, Bailey	165
gregarius, Bréb.	180	cuspidatum	219	Thureti	219
Aptogonum	63, 196	*Cylindrocystis*, Ktz.	153	*Trabecula*, Ehr.	157
Baileyi	208	*Cylindrus*, Ehr.	150	*Trabecula*, Bailey	155
Desmidium	64	decussatum	220	*trabeculoides*, Corda	158
Arthrodesmus	117, 200	Dianæ	168	*truncatum*, Bréb.	156
acutus, Ehr.	191	didymotocum	168	*truncatum*, Kütz.	156
acutus, Bailey	193	*Digitus*, Ehr.	151	turgidulum	221
convergens	118	*dilatatum*, Kütz.	170	turgidum	165
cylindricus, Ehr.	57	*doliolatum*, Bréb.	170	uncinatum	220
Incus	118	Ehrenbergii	166	Venus	220
minutus	213	*elongatum*, Bréb.	173	*verrucosum*, Bailey	218
octocornis, Ehr.	116	*falcatum*, Menegh.	180	*Conferva.*	
pectinatus, Ehr.	191	gracile	221	*dissiliens*, Smith	51
quadricaudatus, Ehr.	190	*granulatum*, Bréb.	147	*mucosa*, Mertens	53
? truncatus	213	*gregarium*, Menegh.	180	Cosmarium	91, 198
Bacillaria.		inæquale	221	aculeatum, Menegh.	113
Lunula, Schrank	163	intermedium	171	amœnum	102
Bambusina.		Jenneri	167	*antilopæum*, Bréb.	114
Brebissonii, Kütz.	59	juncidum	172	*armatum*, Bréb.	112
Binatella.		*læve*, Kütz.	146	attenuatum	110
aculeata, Bréb.	113	*lamellosum*, Bréb.	151	*binale*, Menegh.	90
bacillaris, Bréb.	214	lanceolatum	164	bioculatum	95
hispida, Bréb.	127	Leibleinii	167	*bioculatum*, Menegh.	96
muricata, Bréb.	126	lineatum	173	biretum	102
muticum, Bréb.	125	*lineatum*, Bailey	220	Botrytis	99
tricornis, Bréb.	134	Lunula	163	Brebissonii	100
tricuspidata, Bréb.	122	*Lunula*, Kütz.	166	Broomeii	103
tumida, Bréb.	126	*Lunula*, Leibl.	167	cælatum	103
Closterium	159, 204	*margaritaceum*, Ehr.	149	commissurale	105
acerosum	164	*monile*, Kütz.	145	connatum	108
acuminatum, Kütz.	220	moniliferum	166	conspersum	101
Acus, Nitzsch	175	*nodosum*, Bailey	218	*crassum*, Bréb.	81
acutum	177	obtusangulum	221	crenatum	96
Amblyonema	220	quadrangulare	221	cristatum	105
angustatum	172	Ralfsii	174	Cucumis	93
arcuatum	219	rostratum	175	*Cucumis*, Ralfs	93

INDEX.

	Page
Cucurbita	108
curtum	109
cylindricum	106
deltoides, Corda	99
Didelta, Menegh.	84
elegans, Bréb.	89
gemmatum, Bréb.	87
granatum	96
Incus, Bréb.	118
lagenarium	212
margaritiferum	100
Meneghinii	96
moniliforme	107
oblongum, Bréb.	80
orbiculatum	107
ornatum	104
ovale	98
ovale, Jenner	98
Palangula	212
palmatum, Bréb.	116
Papilio	212
pectinatum, Bréb.	86
Pelta, Corda	211
Phaseolus	106
pyramidatum	94
quadragies-cuspidatum, Corda	209
quadratum	92
Ralfsii	93
sinuosum, Corda	80
Sportella, Bréb.	104
tetraophthalmum	98
Thwaitesii	109
tinctum	95
truncatum, Corda	75
turgidum	110
undulatum	97
ventricosum	212
verrucosum, Menegh.	79
Cylindrocystis.	
Brebissonii, Menegh.	153
truncata, Bréb.	152
Cymbella.	
Hopkirkii, Moore	164
lætevirens, Harvey	153
reniformis, Ag.	100
Desmidium	60, 196
aculeatum, Ehr.	142
apiculosum, Ehr.	126
aptogonum, Bréb.	64
bifidum, Ehr.	215
bispinosum, Corda	xix
Borreri, Ralfs	58
compressum, Ralfs	65
compressum, Corda	57
cylindricum, Greville	57
didymum	207
eustephanum, Ehr.	215
glabrum, Ehr.	217
granulosum, Ehr.	217
hexaceros, Ehr.	134

	Page
limbatum, Chauv.	51
mucosum, Bréb.	51
orbiculare, Ehr.	125
quadrangulare, Kütz.	62
quadrangulatum	62
ramosum, Ehr.	141
senarium, Ehr.	216
Swartzii	61
undulatum	207
vertebratum, Bréb.	65
Diatoma.	
Swartzii, Ag.	61
Didymocladon	144, 203
furcigerus	144
Didymoprium	55, 196
Borreri	58
cylindricum, Ralfs	57
Grevillii	57
Docidium	155, 204
asperum	158
Baculum	158
clavatum	156
constrictum	218
coronatum	217
crenulatum	219
Ehrenbergii	157
minutum	158
nodosum	218
nodulosum	155
truncatum	156
verrucosum	218
verticillatum	218
Echinella.	
acuta, Lyngbye	177
oblonga, Greville	80
rotata, Greville	71
Euastrum	78, 197
aculeatum, Ehr.	209
affine	82
americanum, Ehr.	xix
ampullaceum	83
angulosum, Ehr.	99
ansatum	85
apiculatum, Ehr.	209
armatum, Kütz.	112
binale	90
binale, Kütz.	85
bioculatum, Kütz.	96
Botrytis, Ehr.	99
carinatum, Ehr.	98
circulare	85
convergens, Kütz.	118
cornutum	211
crassum	81
crenatum	211
crenulatum	212
Crux-Melitensis, Ehr.	73
cuneatum	90
Didelta	84
elegans	89

	Page
fasciculatum, Kütz.	114
gemmatum	87
gemmatum, Ralfs	86
hexagonum, Corda	183
hirsutum, Kütz.	127
humerosum	82
incisum, Bréb.	211
Incus, Kütz.	213
insigne	83
integerrimum, Ehr.	93
interstitiale, Kütz.	xix
margaritiferum, Ehr.	100
minutum, Kütz.	213
muricatum, Bailey	210
oblongum	80
octocorne, Kütz.	116
papulosum, Kütz.	87
Pecten, Ehr.	80
pectinatum	86
Pelta	211
Pelta, Ralfs	81
pinnatifidum, Kütz.	77
pinnatum	81
retusum, Kütz.	118
rostratum	88
Rota, Ehr.	70
semiradiatum, Bréb.	75
sinuosum, Lenorm.	85
Sol, Ehr.	72
spinosum, Ralfs	89
sublobatum	91
tetraophthalmum, Kütz.	98
verrucosum	79
Eutomia.	
oblonga, Harvey	80
rotata, Harvey	71
Frustulia.	
acuta, Kütz.	177
subulata, Kütz.	177
Glœoprium.	
dissiliens, Berk.	52
mucosum, Ralfs	53
Goniocystis.	
aculeata, Hassall	142
arachnis, Hassall	137
bifidum, Hassall	131
dilatatum, Hassall	133
gracilis, Hassall	136
hexaceros, Hassall	132
Jenneri, Hassall	129
margaritacea, Hass.	134
mucronata, Hassall	121
muricata, Hassall	127
orbicularis, Hassall	125
paradoxum, Hassall	138
tetracerum, Hassall	137
Helierella.	
Napoleonis, Turpin	186
Boryana, Turpin	187

INDEX.

Heterocarpella.
  binalis, Turpin...... 90
  bioculata, Bréb. ... 95
  Botrytis, Bory ... 99
  Didelta, Turpin ... 84
  tetraophthalma,
    Kütz. 98
  Ursinella, Kütz.... 100
Holocystis.
  oscitans, Hassall... 76
Hyalotheca 51, 196
  bambusina, Bréb. ... 59
  cylindrica, Ehr. ... 57
  dissiliens 51
  dubia 207
  Grevillii, Bréb. ... 57
  Mertensii, Bréb.... 53
  mucosa............... 53
  mucosa, Kütz....... 52
Isthmia.
  filiformis, Menegh. 209
  vertebrata, Menegh. 65
Isthmosira.
  filiformis, Kütz. ... 209
  vertebrata, Kütz... 66
Lunulina.
  monilifera, Bory... 166
  vulgaris, Bory...... 163
Micrasterias 68, 197
  americana xix
  angulosa, Ehr....... 187
  apiculata 209
  Baileyi 211
  Boryana, Ehr. ... 187
  Boryi, Kütz. 187
  Coronula, Ehr. ... 185
  crenata 75
  Crux-Melitensis ... 73
  denticulata 70
  elliptica, Ehr. 188
  falcata, Corda...... 180
  fimbriata 71
  foliacea 210
  furcata................ 73
  heptactis, Ehr...... 183
  incisa 211
  Jenneri 76
  margaritifera,Bréb. 100
  Melitensis, Ralfs... 73
  Melitensis, Menegh. 74
  morsa, Ralfs 74
  muricata 210
  Napoleonis, Kütz. . 186
  octocornis, Menegh. 116
  oscitans 76
  papillifera 72
  pinnatifida 77
  quadragies - cuspidata 209
  radiata, Hassall ... 73
  radiosa 72

Rota, Menegh. ... 70
  rotata 71
  Rotula, Ehr. 183
  Selenæa, Kütz. ... 184
  simplex, Kütz. ... 185
  sinuata, Bréb. ... 80
  Staurastrum, Kütz. 138
  tetracera, Kütz. ... 137
  Tetras, Ehr. 182
  Torreyi 210
  tricera, Kütz...... 137
  tricyclia, Ehr. ... 186
  truncata 75
Mulleria.
  Lunula, Leclerc . . 163
Odontella.
  Desmidium, Ehr.... 64
  filiformis, Ehr. ... 209
  tridentata, Bailey . 208
  unidentata, Ehr.... 65
Palmella.
  cylindrospora,
    Bréb. 153
Pediastrum 180, 205
  angulosum 187
  biradiatum 183
  Boryanum 187
  Boryanum, Ralfs... 188
  constrictum,Hassall 188
  cruciatum, Kütz.... 185
  elegans, Hassall ... 184
  ellipticum 188
  excavatum, Hassall 186
  granulatum 186
  Heptactis............ 183
  Hexactis, Hassall . 186
  lunare, Hassall ... 184
  Napoleonis 186
  Napoleonis, Ralfs . 184
  Napoleonis,Hassall 185
  pertusum 185
  Rotula, Kütz....... 183
  Selenæa 184
  simplex ....:... 185
  simplex, Hassall ... 185
  Tetras 182
  tricyclium, Hassall 186
  vagum, Kütz. 188
Penium 148, 203
  Brebissonii 153
  closterioides 152
  curtum, Bréb. 109
  Cylindrus 150
  Digitus 150
  interruptum 151
  Jenneri 153
  margaritaceum ... 149
  truncatum 152
Pentasterias.
  margaritacea, Ehr. 134
  minor, Kütz. 217

Phycastrum.
  aculeatum, Kütz.... 142
  apiculosum, Kütz. . 126
  bifidum, Kütz....... 215
  cuspidatum, Kütz. 122
  dilatatum, Kütz.... 133
  furcigerum, Kütz. 144
  glabrum, Kütz. ... 217
  granulosum, Kütz. 217
  margaritaceum,
    Kütz. 134
  minus, Kütz. 217
  monticulosum,Kütz. 130
  orbiculare, Kütz... 125
  paradoxum, Kütz. 138
  tricorne, Kütz. ... 135
Pithiscus.
  angulosus, Kütz.... 93
Pleurosicyos.
  myriopodus, Corda 151
Polysolenia.
  Closterium, Ehr.... 151
Rhaphidium.
  duplex, Kütz. 193
  fasciculatum, Kütz. 180
Scenedesmus ... 189, 206
  acutus 191
  antennatus 222
  caudatus, Corda ... 190
  dimorphus 191
  duplex 193
  fusiformis,Menegh. 191
  Leibleinii, Kütz.... 190
  longus, Meyen ... 190
  magnus, Meyen ... 190
  obliquus 192
  obtusus 193
  octoalternus, Kütz. 193
  pectinatus, Meyen 191
  quadralternus,Kütz.193
  quadricauda 190
  quadricaudatus,Ehr.190
  triseriatus, Menegh. 192
Schistochilum.
  excavatum, Ralfs... 67
  unidentatum, Ralfs 66
Selenæa.
  orbicularis, Nitzsch 184
Sphærozosma ... 65, 196
  elegans, Corda ... 65
  excavatum 67
  filiforme 209
  lamelliferum 208
  pulchrum 209
  unidentatum, Ralfs 66
  vertebratum 65
Spirotænia 178, 205
  condensata 179
  obscura 179
Staurastrum ... 119, 201
  aculeatum 142

	Page		Page		Page
aculeatum, Ralfs	142	minus	217	*Tessarthra.*	
alternans	132	monticulosum	130	*filiformis*, Ehr.	209
Arachne	136	*mucronatum*, Ralfs	121	*moniliformis*, Ehr.	107
aristiferum	123	muricatum	126	*Tessarthronia.*	
articulatum	216	*muricatum*, Ralfs	127	*moniliformis*, Turp.	107
asperum	139	muticum	125	Tetmemorus	145, 203
Avicula	140	*octocorne*, Ralfs	116	Brebissonii	145
bacillare	214	orbiculare	125	granulatus	147
bifidum	215	*orbiculare*, Menegh.	126	lævis	146
bifidum, Ralfs	131	paradoxum	138	*Triploceras.*	
brachiatum	131	*paradoxum*, Ehr.	137	*verticillatum*, Bailey	218
brevispina	124	pileolatum	215	*Ursinella.*	
Capitulum	214	polymorphum	135	*margaritifera*, Turp.	100
controversum	141	punctulatum	133	*Vibrio.*	
convergens, Menegh.	118	pungens	130	*acerosus*, Schrank	164
coronatum	216	pygmæum	213	*Lunula*, Müller	163
crenatum	215	quadrangulare	128	Xanthidium	111, 200
cuspidatum	122	rugulosum	214	aculeatum	113
cyrtocerum	139	scabrum	214	*aculeatum*, Bréb.	113
dejectum	121	senarium	216	armatum	112
Dickiei	123	sexcostatum	129	*articulatum*, Corda	216
dilatatum	133	spinosum	143	Artiscon	212
echinatum	215	spongiosum	141	*bisenarium*, Ehr.	113
Ehrenbergii	216	teliferum	128	Brebissonii	113
enorme	140	tetracerum	137	*coronatum*, Ehr.	216
eustephanum	215	tricorne	134	cristatum	115
furcigerum, Bréb.	144	*tricorne*, Ralfs	132	*deltoideum*, Corda	126
glabrum	217	*trilobum*, Menegh.	125	? *difforme*, Ehr.	180
globulatum	217	tumidum	126	*Ehrenbergii*, Corda	216
gracile	136	vestitum	143	fasciculatum	114
granulosum	217	*Stauroceras.*		furcatum	213
hirsutum	127	*Acus*, Kütz.	175	*furcatum*, Ralfs	112
Hystrix	128	*acutum*, Bréb.	177	*hirsutum*, Ehr.	127
Incus, Menegh.	118	*gracile*, Bréb.	221	octocorne	116
Jenneri, Ralfs	129	*Ralfsii*, Bréb.	174	*pilosum*, Ehr.	127
læve	131	*setaceum*, Bréb.	176	*polygonum*, Hassall	114
lunatum	124	*subulatum*, Kütz.	176	*uncinatum*, Bréb.	115
margaritaceum	134	*subulatum*, Bréb.	177		

THE END.

HYALOTHECA. Tab. 1.

1 H. dissiliens. 2 H. mucosa.

E. Jenner, del. W. Willis sc.

DIDYMOPRIUM. Tab. II.

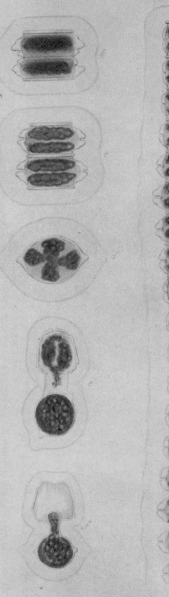

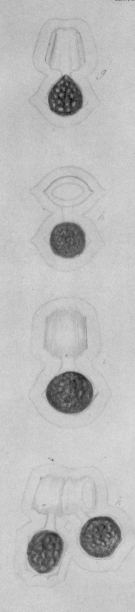

D. Grevillii

DIDYMOPRIUM. Tab. III

D. Borreri.

DESMIDIUM. Tab. IV.

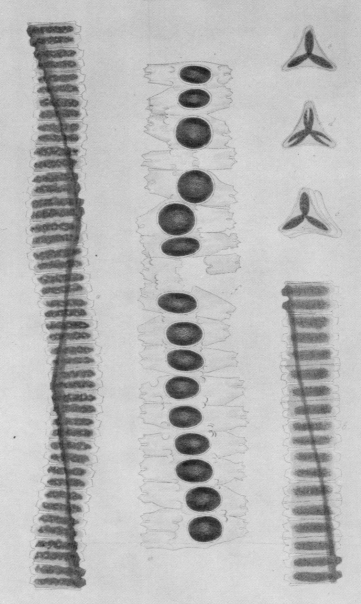

D. Swartzii.

DESMIDIUM. Tab. V.

D. quadrangulatum.

E. Jenner, del. W. Willis, sc.

SPHÆROZOSMA. Tab. VI.

1. S. vertebratum. 2. S. excavatum.

MICRASTERIAS. Tab. VII.

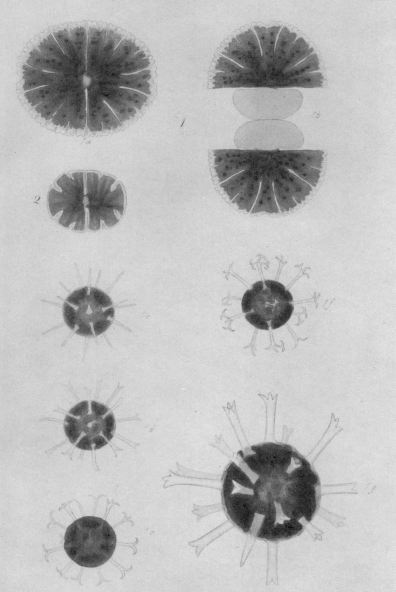

1 M. denticulata.   2 M. crenata.

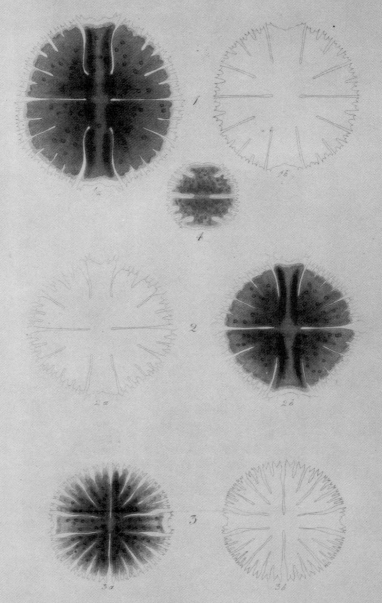

1 M. rotata. 2 M. fimbriata. 3 M. radiosa. 4 M. truncata.

MICRASTERIAS. Tab IX

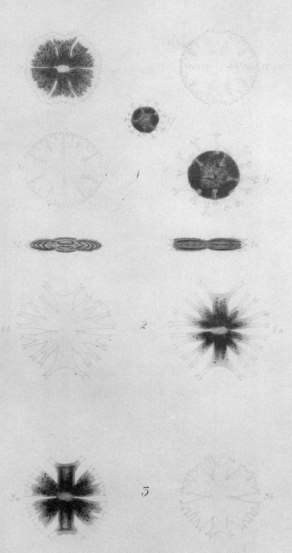

1 M. papillifera. 2 M. furcata. 3 M. Crux Melitensis.

MICRASTERIAS. Tab. X.

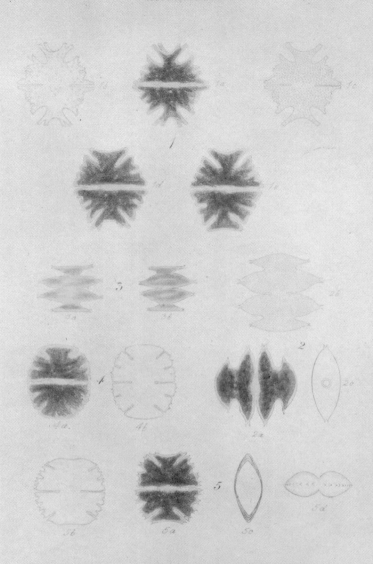

1 *M. mersa.* 2 *M. oscitans.* 3 *M. pinnatifida.*
4 *M. crenata.* 5 *M. truncata.*

MICRASTERIAS — EUASTRUM. Tab. XI.

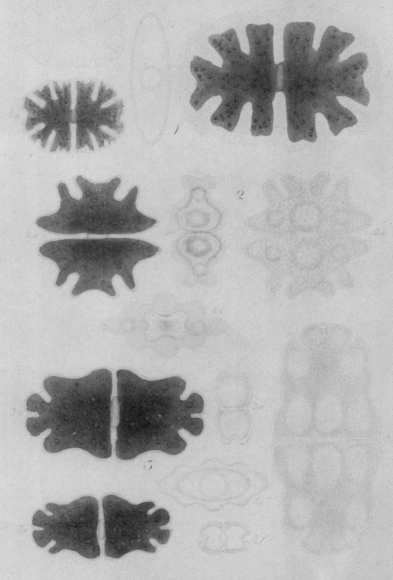

1. M. Jenneri. 2. E. verrucosum. 3. E. crassum.

EUASTRUM. Tab. XII.

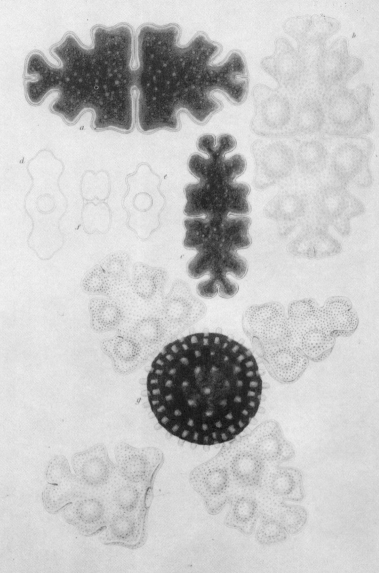

E. oblongum

EUASTRUM. Tab. XIII.

1 E. pinnatum. 2 E. humerosum. 3 E. affine. 4 E. ampullaceum.
5 E. circulare. 6 E. insigne.

EUASTRUM. Tab. XIV.

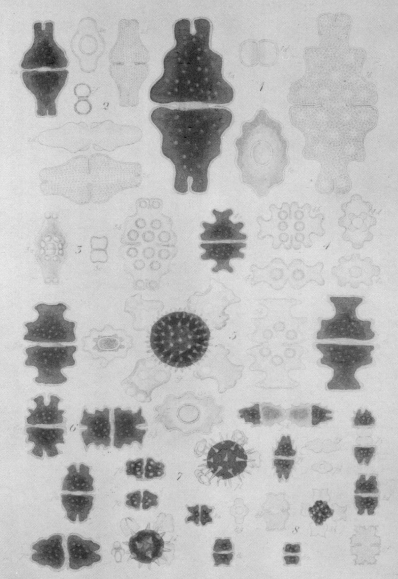

1 E. Didelta. 2 E. ansatum. 3 E. circulare. 4 E. gemmatum.
5 E. pectinatum. 6 E. rostratum. 7 E. elegans. 8 E. binale.

COSMARIUM.   Tab XV.

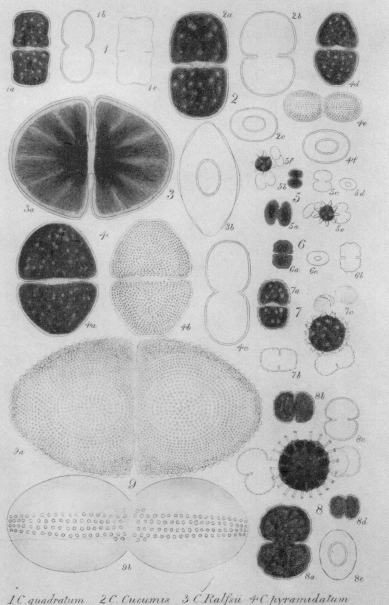

1. C. quadratum   2. C. Cucumis   3. C. Ralfsii   4. C. pyramidatum
5. C. bioculatum   6. C. Meneghini   7. C. crenatum   8. C. undulatum
9. C. ovale

E. Jenner del.   B. Hillmer sc.

COSMARIUM. Tab. XVI.

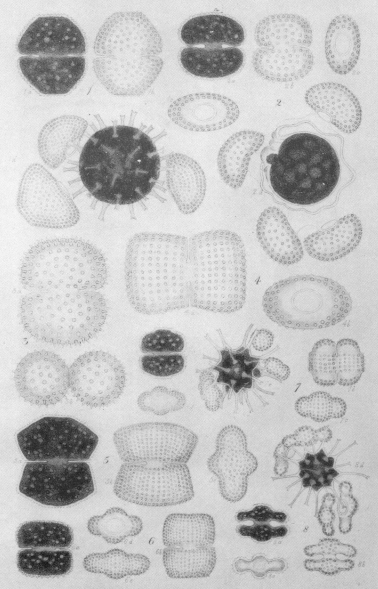

1 C. Botrytis. 2 C. margaritiferum. 3 C. Brebissonii. 4 C. conspersum.
5 C. biretum. 6 C. Broomeii. 7 C. ornatum. 8 C. commissurale.

COSMARIUM

Tab XVII

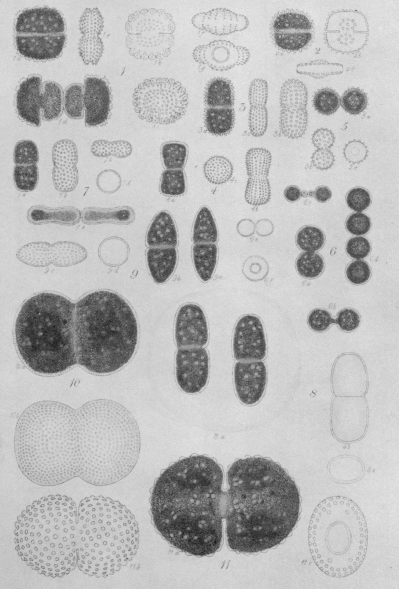

1 C. cælatum. 2 C. cristatum. 3 C. amœnum. 4 C. cylindricum.
5 C. orbiculatum. 6 C. moniliforme. 7 C. cucurbita. 8 C. Thwaitesii.
9 C. attenuatum. 10 C. connatum. 11 C. tetraophthalmum.

E. Jenner, del. W. Willis, sc.

XANTHIDIUM.

Tab. XVIII.

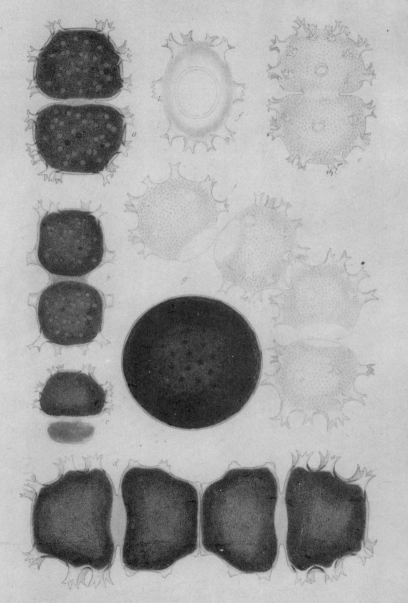

X. armatum.

XANTHIDIUM. Tab. XIX

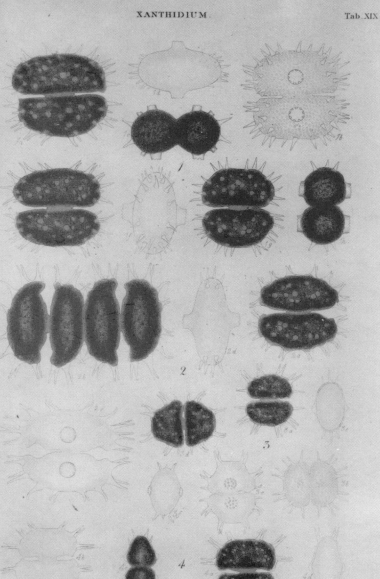

1. X. aculeatum. 2. X. Brebissonii. 3. X. cristatum. 4. X. fasciculatum.

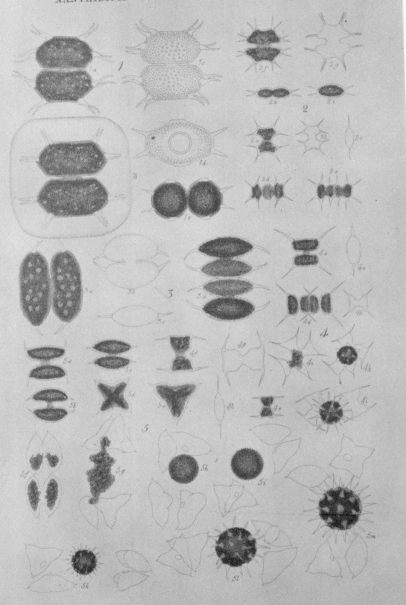

1. X. fasciculatum. 2. X. octocorne. 3. A. convergens. 4. A. Incus. 5. S. dejectum.

STAURASTRUM. Tab. XXI.

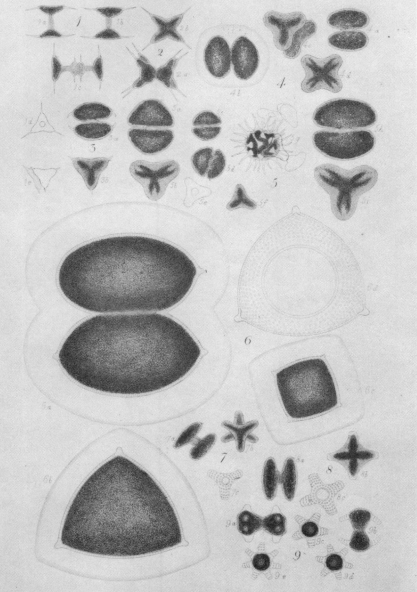

1 S. cuspidatum. 2 S. aristiferum. 3 S. Dickiei. 4 S. muticum.
5 S. orbiculare. 6 S. tumidum. 7 S. alternans. 8 S. dilatatum.
9 S. margaritaceum.

E. Jenner, del. W. Willis, sc.

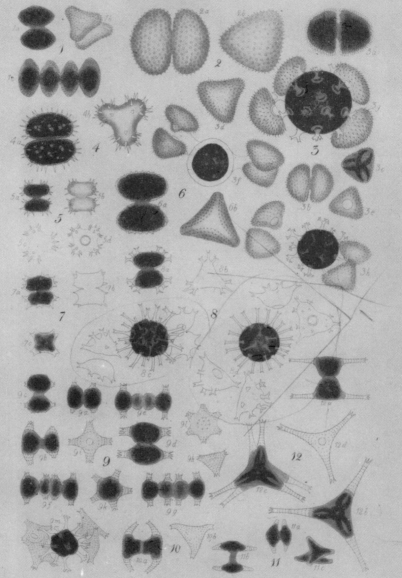

1 *S. punctulatum*. 2 *S. muricatum*. 3 *S. hirsutum*. 4 *S. teliferum*.
5 *S. Hystrix*. 6 *S. asperum*. 7 *S. quadrangulare*. 8 *S. spinosum*.
9 *S. polymorphum*. 10 *S. cyrtocerum*. 11 *S. tricorne*. 12 *S. gracile*.

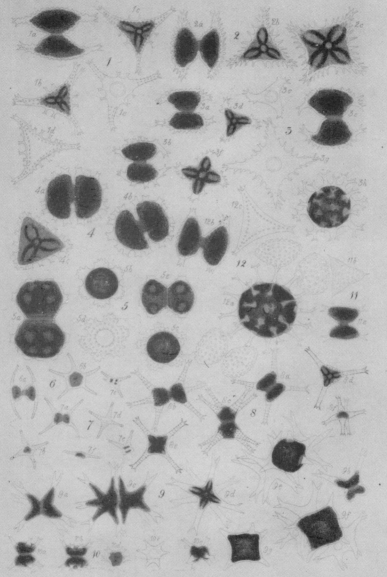

STAURASTRUM Tab. XXIII

1 S. vestitum. 2 S. aculeatum. 3 S. controversum. 4 S. spongiosum.
5 S. sexcostatum. 6 S. Arachne. 7 S. tetracerum. 8 S. paradoxum.
9 S. brachiatum. 10 S. lave. 11 S. Avicula. 12 S. asperum.

TETMEMORUS. Tab. XXIV.

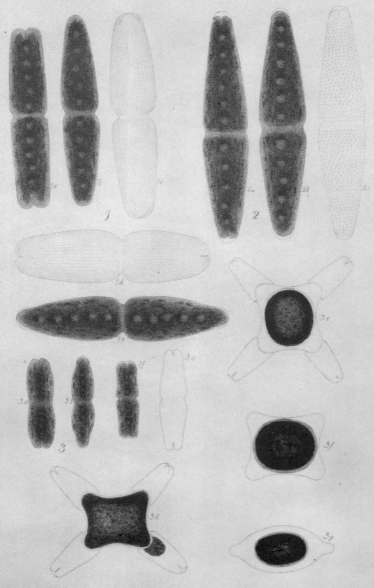

1 T. Brebissonii. 2 T. granulatus. 3 T. lævis.

E. Jenner del. W. Willis sc.

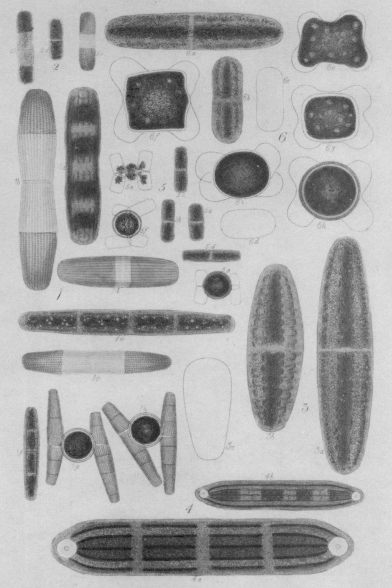

1 *P. margaritaceum*. 2 *P. Cylindrus*. 3 *P. Digitus*. 4 *P. interruptum*.
5 *P. truncatum*. 6 *P. Brebissonii*.

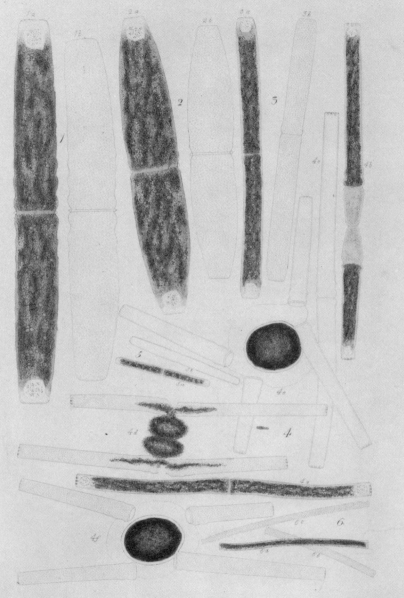

1. D. nodulosum. 2. D. truncatum. 3. D. clavatum. 4. D. Ehrenbergii. 5. D. minutum. 6. D. asperum.

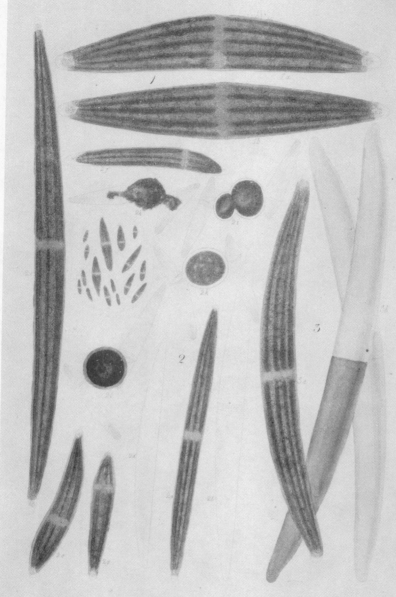

1. *C. Lunula.* 2. *C. acerosum.* 3. *C. turgidum.*

CLOSTERIUM. Tab. XXVIII

1. C. lanceolatum. 2. C. Ehrenbergii. 3. C. moniliferum. 4. C. Leibleinii.
5. C. Dianæ. 6. C. Jenneri. 7. C. didymotocum.

E. Jenner, del. W. Willis, sc.

CLOSTERIUM Tab. XXIX

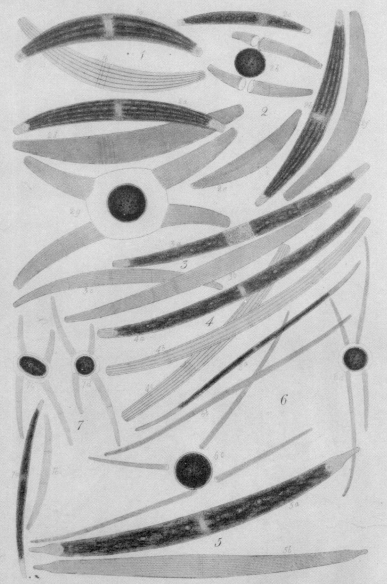

1 C. costatum. 2 C. striolatum. 3 C. intermedium. 4 C. angustatum.
5 C. attenuatum. 6 C. juncidum. 7 C. juncidum β.

E. Jenner del.   W. Willis sc.

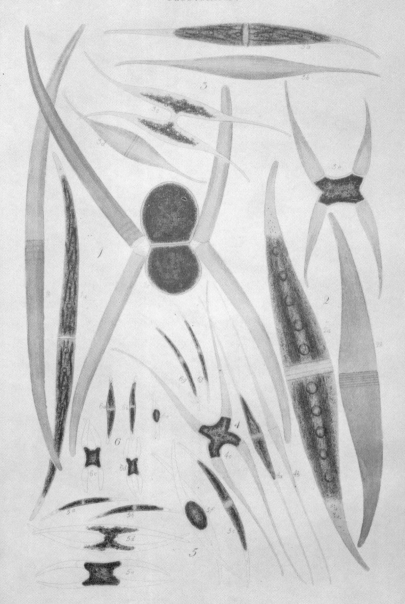

1 C. lineatum. 2 C. Ralfsii. 3 C. rostratum. 4 C. setaceum.
5 C. acutum. 6 C. cornu.

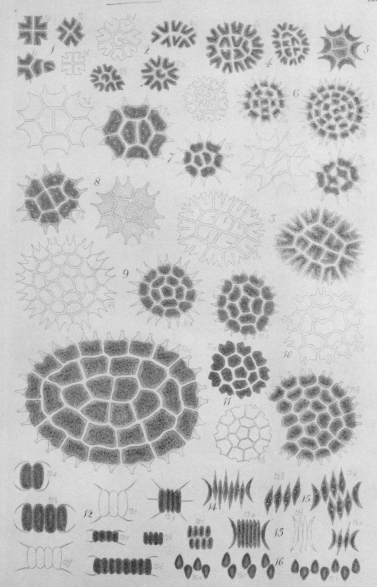

PEDIASTRUM. SCENEDESMUS. Tab. XXXI.

1 P. Tetras. 2 P. Heptactis. 3 P. biradiatum. 4 P. biradiatum β.
5 P. Selenæa. 6 P. pertusum. 7 P. Napoleonis. 8 P. granulatum.
9 P. Boryanum. 10 P. ellipticum. 11 P. angulosum. 12 S. quadricauda.
13 S. dimorphus. 14 S. acutus. 15 S. obliquus. 16 S. obtusus.

E. Jenner, del. W. Willis, sc.

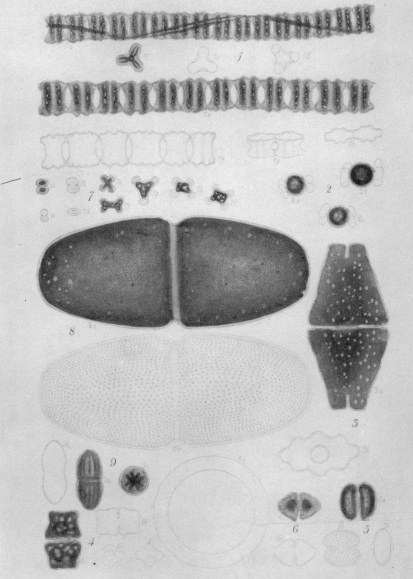

1. A. Desmidium. 2. S. vertebratum. 3. E. cuneatum. 4. E. sublobatum.
5. C. Phaseolus. 6. C. granatum. 7. C. tinctum. 8. C. turgidum.
9. C. curtum.

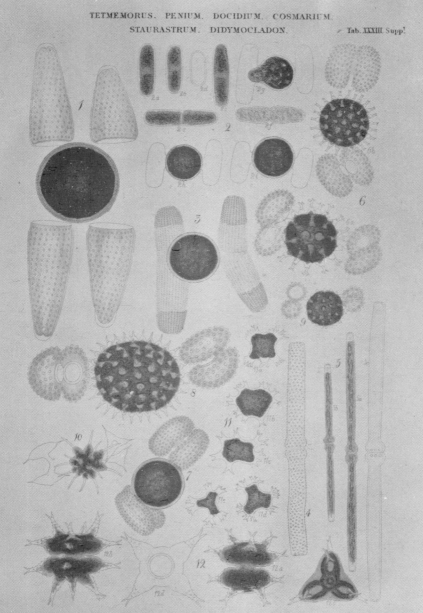

TETMEMORUS. PENIUM. DOCIDIUM. COSMARIUM.
STAURASTRUM. DIDYMOCLADON. Tab. XXXIII Suppl.

1 T. granulatus. 2 P. Jenneri. 3 P. margaritaceum, β. 4 D. Ehrenbergii, β.
5 D. baculum. 6 C. margaritiferum. 7 C. Broomei. 8 C. tetraophthalmum.
9 C. orbiculare. 10 S. cuspidatum. 11 S. enorme. 12 D. furcigerus.

E. Jenner, del. W. Willis, sc.

APTOGONUM. SPHÆROZOSMA. MICRASTERIAS. DOCIDIUM.
CLOSTERIUM. EUASTRUM. ARTHRODESMUS. HYALOTHECA.
STAURASTRUM. SCENEDESMUS.

Tab. XXIV Suppl.

1. A. Baileyi. 2 Sph. pulchrum. 3. M. foliacea. 4. M. Baileyi. 5. M. Torreyi. 6. D. coronatum.
7. D. constrictum. 8. D. nodosum. 9. D. verticillatum. 10. C. tenerrimum. 11. C. cuspidatum.
12. C. Venus. 13. E. cornutum. 14. E. crenatum. 15. A. minutus. 16. H. dubia. 17. S. crenatum.
18. S. maior. 19. S. rugulosum. 20. S. scabrum. 21. S. bacillare. 22. S. pileolatum.
23. S. globulatum. 24. S. echinatum. 25. S. capitulum. 26. S. pygmæum. 27. Sc. antennatus.

W. Willis sc.